Lymphatic System - From Human Anatomy to Clinical Practice

Edited by Gaia Favero and Luca Facchetti

Published in London, United Kingdom

Lymphatic System - From Human Anatomy to Clinical Practice
http://dx.doi.org/10.5772/intechopen.111139
Edited by Gaia Favero and Luca Facchetti

Contributors
Luca Facchetti, Gaia Favero, Samia Hassan Rizk, Ankita Tandon, Kumari Sandhya, Narendra Nath Singh, Ahmed S. Elhamshary, Mostafa I. Ammar, Eslam Farid Abu Shady, Ahmed Elnaggar, Robert M. Molchanov, Oleg B. Blyuss, Ruslan V. Duka, Marzieh Norouzian, Soghra Abdi

Notice
Statements and opinions expressed in the chapters are these of the individual contributors and not necessarily those of the editors or publisher. No responsibility is accepted for the accuracy of information contained in the published chapters. The publisher assumes no responsibility for any damage or injury to persons or property arising out of the use of any materials, instructions, methods or ideas contained in the book.

First published in London, United Kingdom, 2024 by IntechOpen
IntechOpen is the global imprint of INTECHOPEN LIMITED, registered in England and Wales, registration number: 11086078, 5 Princes Gate Court, London, SW7 2QJ, United Kingdom

British Library Cataloguing-in-Publication Data
A catalogue record for this book is available from the British Library

Additional hard and PDF copies can be obtained from orders@intechopen.com

Lymphatic System - From Human Anatomy to Clinical Practice
Edited by Gaia Favero and Luca Facchetti
p. cm.
Print ISBN 978-0-85466-431-3
Online ISBN 978-0-85466-430-6
eBook (PDF) ISBN 978-0-85466-432-0

Meet the editors

Dr. Gaia Favero is a prominent scientist in the field of life sciences. She is currently a researcher for the Scientific-Disciplinary Sector BIO/16 Human Anatomy at the Anatomy and Pathophysiology Division, Department of Clinical and Experimental Sciences, University of Brescia, Italy. Dr. Favero is involved in several crucial topics on morphology, anatomy, and molecular medicine. She focuses her research on aging-related morphological dysfunctions as the prelude to various pathophysiological processes. The central hypothesis is that natural antioxidants may act as molecular "switches" that modulate cells, tissues, and organs by suppressing, at various levels, oxidative stress and inflammatory signaling cascades. This research approach represents powerful tools for developing innovative preventive strategies and identifying novel prognostic biomarkers for several diseases. During her research residency program, Dr. Favero completed an internship at the Department of Physiology and Pharmacology, Health Science Campus, University of Toledo Medical Center, USA). She has about 150 scientific publications to her credit.

Dr. Luca Facchetti, MD, EdiR, is a radiology physician in the Department of Radiology, 1st division, Spedali Civili di Brescia, Italy. He subspecializes in breast radiology and intervention. During his residency program, he completed a research fellowship at the Radiology and Biomedical Imaging Department, University of California San Francisco, USA, and a clinical fellowship in breast imaging at the South General Hospital Breast Center, Sweden. He is a member of the Italian and Norwegian boards of medicine and radiology. Dr. Facchetti has twenty-seven scientific publications to his credit.

Contents

Preface

The lymphatic system has a fundamental role in interstitial fluid drainage, but it also has important immune functions. The lymphatic system carries lymph through a network of lymphatic channels (capillaries and vessels), then the lymph is filtered by lymph nodes and returned to the bloodstream, where it is eventually eliminated.

Notably, the lymphatic system also has a critical role in a clinical context: the inflammation of lymphatic vessels and lymph nodes may represent an indicator of pathology. The lymphatic and vascular systems have numerous connections and tumor cell metastasis may pass from one circulatory system to the other.

This book presents a comprehensive overview of the lymphatic system, including but not limited to the human anatomy of the lymphatic system and its involvement in health and disease conditions. It clarifies the anatomy of the human lymphatic system to help identify and develop novel therapeutic approaches focused on lymphatic system response.

The editors would like to thank all the authors who contributed to the success of this book and the staff at IntechOpen for their valuable and constant support.

Gaia Favero, Ph.D.
Anatomy and Physiopathology Division,
Department of Clinical and Experimental Sciences,
University of Brescia,
Brescia, Italy

Luca Facchetti, MD
Department of Radiology 1st Division,
Territorial Social Health Authority (ASST) of the Civil Hospitals of Brescia,
Brescia, Italy

Chapter 1

Introductory Chapter: Lymphatic System Human Anatomy

Luca Facchetti and Gaia Favero

1. Introduction

1.1 Lymphatic system human anatomy

The lymphatic system modulates the interstitial fluid volume through a one-way transport system. The residual interstitial fluid is a carriage from the soft tissue interstitial space into the venous circulation through specific lympho-venous connections [1, 2]. Along with the excess interstitial fluid, redundant proteins and "waste" are transported back to the bloodstream by the lymphatic system. In detail, the lymphatic system *via* a network of lymphatic channels transports through lymph nodes the interstitial fluid, which is defined as lymph when it is inside the lymphatic channels network, and discharges it into the blood circulation [3–5]. Lymph nodes filter the interstitial flow and break down bacteria, viruses, and other cells and molecules. The lymphatic system is also important in immune surveillance defending the body against microorganisms and foreign particles. The lymphatic system encourages the immune response [2, 6, 7]. The lymphatic system is, therefore, strictly correlated to the circulatory system and immune system, but not only. In fact, the lymphatic system has also an important role in the absorption of fat-soluble vitamins and fatty substances at the gut level, through gastrointestinal tract's specific lymphatic capillaries called lacteals, so dietary fat is transported into the venous circulation [1]. In addition, lymphatic vessels were recently identified in the brain meninges and the meningeal lymphatic network is fundamental for the removal of toxins and also in draining cerebrospinal fluid and immune cells from the central nervous system to the peripheral lymphatic system [3, 8].

The lymphatic system is a highly complex system with a variable structure and function between anatomical sites and between species. In humans, the lymphatic system includes lymphatic capillaries, lymphatic vessels (afferent and efferent), lymph nodes, and various lymphoid organs (such as the thymus and spleen) [2, 6, 9].

In this introductory chapter, we will focus our attention mainly on the description of the hierarchy lymphatic channels network (lymphatic capillaries and lymphatic vessels).

The organization of lymphatic networks within various organs depends on the functional demands of the organ itself, leading to both common and unique morphological features of the lymph-venous connections and lymphatic channel network [3, 10]. Interstitial fluid comes out of the blood capillary walls due to heart and/or cell osmotic pressure and enters the lymphatic system through small and blind-ended lymphatic capillaries. These capillaries, defined as initial lymphatics, form

IntechOpen

a mesh-like network and gradually increase their diameter becoming pre-collector vessels, collector vessels, lymphatic trunks, and finally ducts. When soft tissue interstitial pressure increases, the interstitial fluid enters into the lymphatic capillaries through openings in the endothelial layer; whereas, when the pressure inside the lymphatic capillaries rises, the interstitial fluid entering flow is stopped. Lymphatic capillaries are tiny, thin-walled, and blind-end channels that present a larger diameter with respect to blood capillaries. In addition, the lymphatic capillaries are dissipated among blood capillaries to facilitate interstitial fluid collection by the lymphatic capillary network. The lymphatic capillaries endothelial cells overlap but shift to open the capillary wall when interstitial fluid pressure is greater than intra-capillary pressure so permitting interstitial fluid, lymphocytes, bacteria, cellular debris, plasma proteins, and other cells to enter the lymphatic capillaries [3, 6, 11]. The interstitial fluid inside the lymphatic channels network is defined as lymph. Lymph is composed of interstitial fluid with variable amounts of lymphocytes, monocytes, plasma proteins, and other cells. Lymph formation is organ-dependent and it is correlated to the various organs/tissues morphostructural properties [2, 6].

The lymphatic capillaries form large networks of channels called lymphatic plexuses and converge to form larger lymphatic vessels. Collecting vessels are further divided into afferent (pre-nodal) and efferent (post-nodal) vessels depending on their location relative to lymph nodes. Afferent lymphatic vessels transport the unfiltered lymph from tissues to the lymph nodes and efferent lymphatic vessels convey filtered lymph from lymph nodes to subsequent lymph nodes or into the venous system [6]. Lymph flow generally occurs against a pressure gradient and therefore requires both extrinsic forces, such as skeletal muscle movement and arterial pulsations, and intrinsic forces exerted by lymphatic vessels. In fact, the lymph is pumped slowly by the contraction of the lymphatic vessels [2, 3, 12]. To prevent lymph flow backward, collecting lymphatic vessels and larger lymphatic vessels presented a series of one-way valves; notably, the one-way valves are not present in the lymphatic capillaries. These lymphatic valves help the advancement of lymph flow through the lymphatic vessels.

The anatomical structure of each component of the lymphatic vessel network and its surroundings contribute to its function. **Figure 1** reported a schematic representation of the lymphatic channel network.

The lymphatic channels gradually increased their diameter becoming finally the main lympho-venous connection: the thoracic duct and the right lymphatic duct. The right lymphatic duct is responsible for draining the lymph from the upper right quadrant of the body (the right side of the head, neck, thorax and the right upper limb) into the venous circulation at the junction between the right subclavian vein and the right internal jugular vein. The right lymphatic duct is formed generally by the convergence of the right bronchomediastinal trunk, jugular trunk, and subclavian trunk [6]. However, it is important to underline that its origin and ending presented a changeable anatomy and morphology.

The thoracic duct, also known as the left lymphatic duct or van Hoorne's canal, drains the lymph of the body except for the territory drained by the right lymphatic duct so it drains lymph from 80% to 90% of the body. The thoracic duct is a thin-walled tubular lymphatic vessel (with 2–6 mm in diameter). The thoracic duct is the largest and longest lymphatic duct in the body. This duct drains lymph at the junction between the left internal jugular vein and the left subclavian vein [1, 2]. The thoracic duct presents a high anatomical variability, but it typically arises in the abdomen as

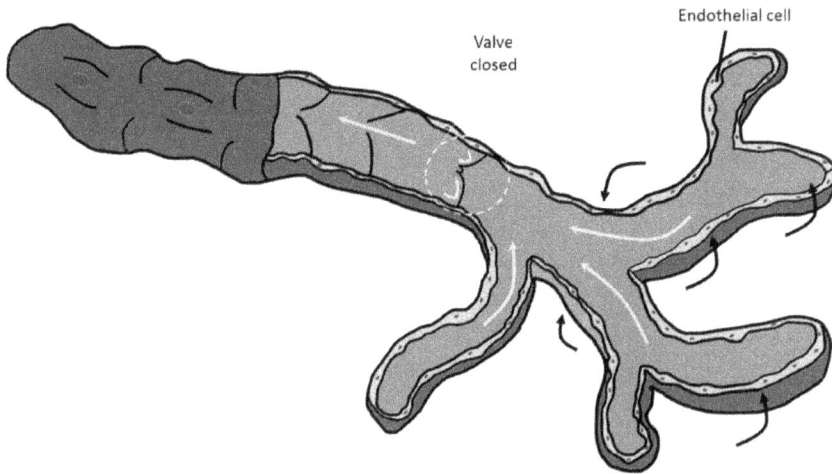

Figure 1.
Schematic representation of the lymphatic channel network showing lymphatic capillaries and vessels. The yellow arrows represent the direction of lymph flow inside the lymphatic channels and the black arrows show the interstitial fluid entering the lymphatic vessels.

cisterna chyli, which is an expanded lymphatic sac that forms at the convergence of the intestinal lymphatic trunk and lumbar lymphatic trunk. The cisterna chyli is at the level of the 12 thoracic vertebrae (T12) [13]. Notably, the cisterna chyli is present in approximately 40–60% of the population and in people without this cisterna the intestinal and lumbar lymphatic trunks communicate directly with the thoracic duct [14]. From the cisterna, the thoracic duct ascends running to the right of the body midline and posterior to the aorta and it enters the thorax *via* the aortic hiatus. The thoracic duct then rises in the thoracic cavity anteriorly and to the right of the vertebral column, between the aorta and azygous vein. At about the level of the fifth thoracic vertebra (T5), the thoracic duct crosses to the left of the vertebral column and posterior to the esophagus. Finally, it ascends vertically and then releases the drained lymph in the venous circulation [6, 15, 16].

There is so a continuous and dynamic exchanging circulation of extracellular fluid passing back and forth from the bloodstream to the tissues and lymphatic system. The occlusion of the lymphatic vessels downstream may promote the opening of new lympho-venous connections, resulting in anatomo-morphological changes relevant in both health and disease states [1, 5, 12].

2. Conclusion

The lymphatic system has a fundamental role in extracellular fluid drainage, but it has also important immune functions. The lymphatic system has also a critical role in a clinical context, because the lymphatic and vascular systems have numerous connections, and tumor cell metastasis may pass from one circulatory system to the other.

The lymphatic system is a highly complex and dynamic system and the lymphatic system of human anatomy is fundamental to modulate pathological changes which are relatively unknown, but fundamental in clinical practice.

Acknowledgements

The authors sincerely thank Dr. Marzia Gianò (Anatomy and Physiopathology Division, University of Brescia, Italy) for her assistance in drawing the figure.

Conflict of interest

The authors declare no conflict of interest.

Author details

Luca Facchetti[1] and Gaia Favero[2,3]*

1 Department of Radiology 1st Division, ASST Spedali Civili of Brescia, Brescia, Italy

2 Anatomy and Physiopathology Division, Department of Clinical and Experimental Sciences, University of Brescia, Brescia, Italy

3 Interdipartimental University Center of Research "Adaption and Regeneration of Tissues and Organs -ARTO", University of Brescia, Brescia, Italy

*Address all correspondence to: gaia.favero@unibs.it

IntechOpen

References

[1] Hsu MC, Itkin M. Lymphatic anatomy. Techniques in Vascular and Interventional Radiology. 2016;**19**:247-254. DOI: 10.1053/j.tvir.2016.10.003

[2] Margaris KN, Black RA. Modelling the lymphatic system: Challenges and opportunities. Journal of the Royal Society Interface. 2012;**9**:601-612. DOI: 10.1098/rsif.2011.0751

[3] Jayathungage Don TD, Safaei S, Maso Talou GD, Russell PS, Phillips ARJ, Reynolds HM. Computational fluid dynamic modeling of the lymphatic system: A review of existing models and future directions. Biomechanics and Modeling in Mechanobiology. 2023. DOI: 10.1007/s10237-023-01780-9 [Epub ahead of print]

[4] Leong SP, Witte MH. Future perspectives and unanswered questions on cancer metastasis and the lymphovascular system. Cancer Treatment and Research. 2007;**135**:293-296. DOI: 10.1007/978-0-387-69219-7_21

[5] Threefoot SA. Gross and microscopic anatomy of the lymphatic vessels and lymphaticovenous communications. Cancer Chemotherapy Reports. 1968;**52**:1-20

[6] Null M, Arbor TC, Agarwal M. Anatomy, lymphatic system. In: StatPearls. In; 2023

[7] Swartz MA, Hubbell JA, Reddy ST. Lymphatic drainage function and its immunological implications: From dendritic cell homing to vaccine design. Seminars in Immunology. 2008;**20**:147-156. DOI: 10.1016/j.smim.2007.11.007

[8] Louveau A, Smirnov I, Keyes TJ, Eccles JD, Rouhani SJ, Peske JD, et al. Structural and functional features of central nervous system lymphatic vessels. Nature. 2015;**523**:337-341. DOI: 10.1038/nature14432. Erratum in: Nature. 2016;533:278

[9] Cao Y, Chen H, Yang J. Neuroanatomy of lymphoid organs: Lessons learned from whole-tissue imaging studies. European Journal of Immunology. 2023;**53**:e2250136. DOI: 10.1002/eji.202250136

[10] Breslin JW, Yang Y, Scallan JP, Sweat RS, Adderley SP, Murfee WL. Lymphatic vessel network structure and physiology. Comprehensive Physiology. 2018;**9**:207-299. DOI: 10.1002/cphy.c180015

[11] Serrano JC, Gillrie MR, Li R, Ishamuddin SH, Moeendarbary E, Kamm RD. Microfluidic-based reconstitution of functional lymphatic microvasculature: Elucidating the role of lymphatics in health and disease. Advanced Science (Weinheim, Baden-Württemberg, Germany). 2023:e2302903. DOI: 10.1002/advs.202302903

[12] Scallan JP, Zawieja SD, Castorena-Gonzalez JA, Davis MJ. Lymphatic pumping: Mechanics, mechanisms and malfunction. The Journal of Physiology. 2016;**594**:5749-5768. DOI: 10.1113/JP272088

[13] Moazzam S, O'Hagan LA, Clarke AR, Itkin M, Phillips ARJ, Windsor JA, et al. The cisterna chyli: A systematic review of definition, prevalence, and anatomy. American Journal of Physiology. Heart and Circulatory Physiology. 2022;**323**(5):H1010-H1018. DOI: 10.1152/ajpheart.00375.2022

[14] Phang K, Bowman M, Phillips A, Windsor J. Review of thoracic duct anatomical variations and clinical implications. Clinical Anatomy. 2014;**27**:637-644. DOI: 10.1002/ca.22337

[15] O'Hagan LA, Windsor JA, Itkin M, Russell PS, Phillips ARJ, Mirjalili SA. The lymphovenous junction of the thoracic duct: A systematic review of its structural and functional anatomy. Lymphatic Research and Biology. 2021;**19**:215-222. DOI: 10.1089/lrb.2020.0010

[16] O'Hagan LA, Windsor JA, Phillips ARJ, Itkin M, Russell PS, Mirjalili SA. Anatomy of the lymphovenous valve of the thoracic duct in humans. Journal of Anatomy. 2020;**236**:1146-1153. DOI: 10.1111/joa.13167

Chapter 2

Bone Marrow Lymphocytes' Development and Dynamics

Samia Hassan Rizk

Abstract

The bone marrow (BM) is an integral part of the immune system that communicates with other immune tissues via the bloodstream but does not have lymphatic vessels. It is the primary site of lymphopoiesis, where B cells and early T-cell progenitors develop, from late fetal life onwards, and a secondary lymphoid organ for B lymphocytes. At the same time, it regulates the function and dynamics of the immune system in a steady state and disease conditions. Activating and inhibitory signals from various marrow elements regulate the traffic of lymphocyte subtypes (B, T, and NK), including direct cell contact and released factors from stromal cells. This chapter is a review of the life cycle and dynamics of lymphoid cells in health and representative immune-associated disorders. Understanding the central bone marrow's role may clarify the pathologic changes and open potential therapeutic channels in some disorders.

Keywords: lymphopoiesis, bone marrow niche, lymphocyte function regulation, bone marrow lymphocyte types, bone marrow immune role in disease

1. Introduction

Both hematopoietic and lymphoid systems arise in the bone marrow from the late fetus and prevail throughout life with a finite control by lineage-specific and broadly acting biological factors and a homeostatic signaling network, allowing sensitive responsiveness to fluctuating body needs [1]. The complex functional structure of the bone marrow exhibits age and context adaptation [2].

The bone marrow (BM) is a central immune organ that does not have lymphatic vessels but communicates with other lymphoid tissues via the bloodstream. It is also a secondary lymphoid organ for mature B cells. It differs from other immune organs in lacking a fixed lymphocyte organization but contains interstitially scattered cells within its parenchyma, sometimes, forming small aggregates, especially in the elderly. In the bone marrow, regulatory cells and short- and long-acting signals orchestrate the immune system and critically underlie the pathogenesis of many reactive and neoplastic conditions [1]. This chapter reviews the bone marrow structure and function as a part of the immune system, including the following three sections:

IntechOpen

1. Bone marrow lymphopoiesis: ontogeny and maturation

2. The milieu and kinetics of bone marrow lymphoid elements

3. Dynamics of BM lymphoid cells in health and disease

2. Bone marrow lymphopoiesis: ontogeny and maturation

The hematopoietic marrow extends throughout the medullary bone cavities of the body from the late fetus along the life span, gradually contracting toward the axial skeleton with age. In these locations, interaction with a complex connective tissue stroma with cellular and extracellular matrix components is critical for supporting and regulating proliferation, maturation, maintenance, senescence, and final destruction of hematopoietic elements.

Stromal elements comprise bone, vasculature, and a network of mesenchymal and reticular cells, critical for bone marrow functioning. Other elements include sympathetic and parasympathetic innervation, adipocytes, resident macrophages [3], neutrophils [4], megakaryocytes [5], T lymphocytes, and dendritic cells (DCs). The extracellular matrix of macromolecules, including fibronectin, vitronectin, collagens (types: I, II, IV, and VI), and proteoglycans, forms an integral part of the hematopoietic stem cell (HSC) microenvironment. Specific HSC niches refer to the arrays of stromal cells, specific locations, soluble molecules, signaling cascade, and gradients, together with the shear stress, temperature, and oxygen tension, which determine the stem cell behavior at any given time [6]. The niche dynamics promote specific HSC properties such as; quiescence, self-renewal, and proliferation. In contrast to other organs, the bone marrow has no clear demarcation by distinct cell density or stromal cell distribution.

Both hematopoietic and lymphoid progenitors have a common origin from pluripotent CD34+ stem cells (multipotent progenitors (MPPs)) in the bone marrow. Lymphoid lineages originate from a common progenitor with the central control of the Ikaros gene; a zinc finger gene encoding DNA-binding protein transcription factor, which plays crucial functions in hematopoiesis and regulation of immune cell development. Along with other transcription factors, it also regulates the expression of other genes influencing the phenotypic characteristics of lymphocytes, including immunoglobulin heavy and light chain gene rearrangement, and cluster of differentiation 3 (CD3) complex antigen receptors.

After an initial commitment, T-cell precursors migrate to the thymus, while B-cell progenitors complete their primary maturation in the bone marrow. **Figure 1** outlines the central and peripheral compartments, and stages of lymphopoiesis [7]. Receptor gene recombinations occur along B- and T-lymphocyte differentiation before emigrating to the peripheral lymphoid tissues or the peripheral blood [1].

2.1 B-cell maturation

2.1.1 Primary B-cell maturation

Primary B-cell maturation proceeds through four main stages; pro-B, pre-B, immature B cells, and mature B cells [8], with an essential requirement for stromal interaction by direct cell contact and released growth factors [9]. Direct contact of pro-B cells with stromal cells occurs via V and LA-4/M1, and then c-kit receptor/

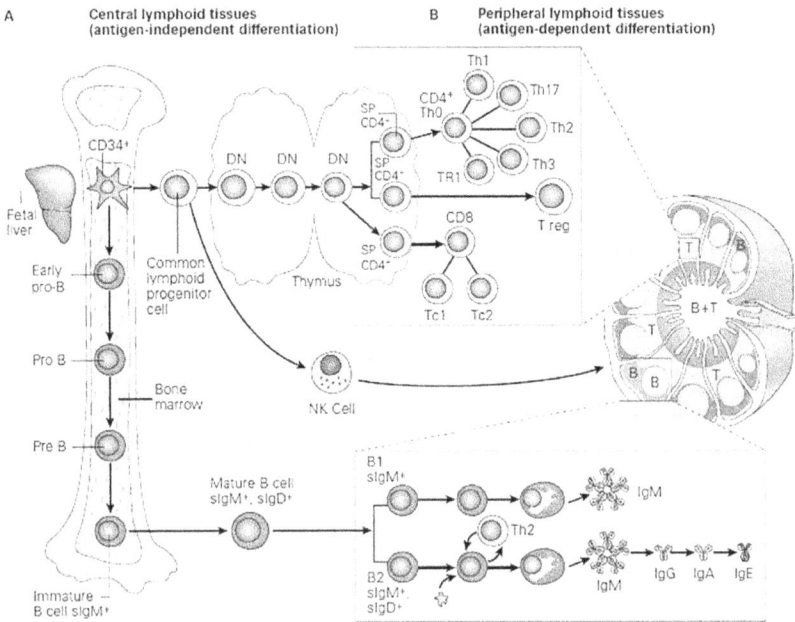

Figure 1.
Schematic representation of the central and peripheral compartments of lymphoid differentiation: Panel a: Development of the immune system from stem cells in bone marrow, and differentiating in central lymphoid tissues (bone marrow and thymus) is independent of antigen contact. Panel B: Migration of cells into peripheral lymphoid tissues (lymph nodes, spleen, and mucosa-associated lymphoid tissues) is antigen-dependent [7].

stromal cells-stem-cell factor interaction initiates B-cell division and interleukin-7 (IL-7) receptor expression. Eventually, the downregulation of adhesion molecules releases the B cells.

The recombination of heavy chain variable gene region segments marks the pre-B cells [10]. The next stage of immature B cells expresses membrane immunoglobulin M (IgM).

B cells recognizing self-antigens undergo apoptosis and the surviving cells differentiate along two subpopulations: B1 cells expressing CD19+, CD5+, and sIgM+, and B2 cells expressing CD19+, sIgM+, and sIgD+. Both subtypes migrate into the peripheral lymphoid organs [10].

2.1.2 Secondary B-cell differentiation

In peripheral lymphoid organs, B-cell differentiation is antigen-driven with regulation by sequentially acting transcription factors that induce initial upregulation of PAX5, IRF8, and BACH2 genes, followed by IRF4, XBP1, and BLIMP1 genes [11]. In parallel with these changes, B cells express stage-specific markers, including cluster of differentiation 27 (CD27), cluster of differentiation 38 (CD38), and cluster of differentiation 138 (CD138), with simultaneous downregulation of the B-cell markers such as cluster of differentiation 19 (CD19) and cluster of differentiation 20 (CD20).

In the germinal centers, signals from the helper T cells activate B cells that proliferate, undergo somatic hypermutation, and finally generate memory B cells and long-lived plasma cells [11].

Although B-cell development is conceptually unidirectional, recent evidence suggests a possible phenotypic "reflex" from immature to pre-B cell subsets, probably through receptor editing, though the existence of multiple differentiation pathways is also possible [12].

2.1.3 The plasma cell stage

The maintenance of plasma cell survival in the bone marrow depends on stromal cell signals, and cytokines, including megakaryocyte proliferation-inducing ligand-APRIL and interleukin-6 (IL-6) [13], eosinophils APRIL and IL-6, [14], and mono-cytes APRIL [15], while the granulocyte colony-stimulating factor (G-CSF) mobilizes plasma cells [16].

2.2 T-cell maturation

After a short bone marrow phase, cytokines and major histocompatibility complex class I or II (MHC-I or II) on thymus epithelium promote the differentiation of T pre-cursor cells into T-central (Tc), regulatory T cells (Treg), and T helper (Th) subpopu-lations, while those with self-specific T-cell receptor (TCR) undergo apoptosis [17].

Bone marrow T cells contribute to immune homeostasis where IL-7, interleukin-15 (IL-15), and tumor necrosis family (TNF) family members activate CD8+ T cells. Priming of naive CD4+ and CD8+ T cells may occur in response to bloodborne anti-gens presented by bone marrow dendritic cells (DCs) and other myeloid elements. Survival of memory CD8+ and CD4+ T cells in the bone marrow requires IL-7 and IL-15, and Major Histocompatibility Complex (MCH) and IL-7, respectively [17].

At least two memory T-cell niches sustain the viability and functionality of CD4+ T cells [17], similar to the hematopoietic stem cell niches [18]. CD45RA memory T cells include CCR7+ (central memory cells) and CCR7 effector memory cells [19], while noncirculating cells constitute "Tissue-resident memory T cells (Trm cells)" [18].

2.3 Natural killer cells

Natural killer (NK) cells develop from bone marrow CD34+ stem cells and undergo similar maturation stages to other lymphoid cells, but lack antigen-specific receptors. They express receptors for activating cellular killing and other receptors recognizing self-MHC alleles that inhibit the killing of normal cells. NK cells have the appearance of large granular lymphocytes with neither T- nor B-cell antigens and do not express cluster of differentiation 16 (CD16) and cluster of differentiation 56 (CD56) (an NK-specific adhesion molecule), helping their identification in the peripheral circulation [1].

2.4 Regulation of bone marrow lymphopoiesis

In the bone marrow, lymphopoiesis shares a common perivascular niche with myelo-poiesis with several niche factors regulating their maturation and functions, including adipocytes [20], regulatory T cells (Tregs) [21], monocytic cells, mesenchymal cells [22], and nerves [23]. External signals also contribute to lymphocyte regulation such as sex steroids. This arrangement allows a versatile hemopoietic response under stress conditions, related to competitive and differential requirements for cytokines such as C-X-C motif chemokine ligand 12 (CXCL12) and stem cell factor (SCF) [24].

Bone marrow stromal cells maintain the survival and function of B lymphocytes, long-lived plasma cells, and memory T cells. They also release inhibitory factors such as transforming growth factor-β (TGF-β) and hepatocyte growth factor (HGF). In steady states, resident, low-migrating mature cluster of differentiation 19 (CD19)-negative plasma cells are the major B-cell component in the bone marrow. During an immune response, CD19+ plasma cells predominate [25]. Reciprocally, activated T cells secrete cytokines that promote the terminal differentiation of myeloid precursors. Unlike hemopoietic cells, plasma cells do not have fixed anatomical niches in the bone marrow [26], but receive supporting signals from many cells, including, CXCL12-producing mesenchymal cells [27], granulocytes, megakaryocytes, and myeloid cells [13]. Additional factors influencing plasma cell kinetics include megakaryocyte proliferation-inducing ligand APRIL and IL-6 [14], eosinophils APRIL and IL-6 [13], and monocytes APRIL [15]. On the other hand, G-CSF mobilizes plasma cells [16].

3. The milieu and kinetics of BM lymphoid elements

The bone marrow hosts a mixture of mature lymphocytes in specific niches, which promotes their long-term survival, including B, T, and NK cells [28]. It harbors about 12% of all lymphoid cells in the human body at any given time. Most T cells are recirculating from peripheral lymphoid tissues including memory CD4$^+$ and CD8$^+$ cells. Plasma cells similarly migrate into the bone marrow where they reside by losing response to CXCL12 and C-X-C motif chemokine ligand 9 (CXCL9) signals [25].

In normal adults, bone marrow lymphocytes range between 10 and 20% of cellularity; the majority are small cells, resembling those in the peripheral blood, and scatter interstitially with occasionally small aggregates, especially in the elderly. They comprise a mixture of B, T, and NK cells with their subtypes. T cells are the predominant lymphocytes in the bone marrow with a T-to-B lymphocyte ratio of 4 to 1. The CD4+ to CD8+ ratio is 1:2; the reverse of that in the peripheral blood.

3.1 The bone marrow lymphocyte mix

3.1.1 Bone marrow, B lymphocytes

The bone marrow is the site for B lymphopoiesis throughout life. Early stages of B-cell maturation prevail during fetal life, whereas mature B cells predominate in adults [26]. In infants and young children, the marrow has a higher percentage of lymphocytes (up to 50% of cellularity), including a proportion of larger lymphocytes with immature morphology (hematogones) that represent lymphoid progenitors and decrease with age [27]. The bone marrow has a higher proportion of IgM- than IgG (immunoglobulin G)-bearing B cells, which is the reverse of the peripheral blood [28]. They also express lower levels of peanut agglutinin (PNA) relative to germinal centers or circulating memory B cells [26].

In the adult bone marrow, most B cells have prior activation in germinal centers of peripheral lymphoid tissues, evidenced by somatic hypermutation, isotypic diversification, and antigen selection. Those expressing IgM lack cluster of differentiation 10 (CD10), lower cluster of differentiation 24 (CD24), and co-expression of immunoglobulin D (IgD), which differentiate them from the *in situ*-generated immature B cells [26].

3.1.2 Bone marrow, plasma cells

Plasma cells are terminal, nonproliferative B cells that develop through T-cell-dependent or -independent pathways [29]. A continuum of plasma cells' maturation stages ranging from the less mature, proliferative short-lived (CD138+ B220+ or plasmablasts), to the mature long-lived PCs (LLPCs) are present. The latter make up more than 50% of total bone marrow plasma cells [29].

Factors maintaining survival of plasma cells in the bone marrow include membrane-bound ligands like cluster of differentiation 80 (CD80) and secreted proteins such as APRIL, interleukin-6 (IL-6), B-cell activating factor (BAFF), and CXCL12 (or stromal cell-derived factor-1 (SDF-1)) [30]. In Giemsa-stained marrow, plasma cells are scattered singly or in small groups representing about 2% of cellularity in normal adults. In biopsy sections, they appear close to capillaries [27]. They have a unique migration pattern with alternating high motility and low-rate migration or arrest, closely linked to their survival [29]. APRIL-secreting cells dynamically coalesce and recruit migrating plasma cells into clusters and release signals that promote the overall motility of plasma cells [31], and enhance their fitness and survival [29]. Other factors enhancing plasma cell motility include CXCL12 and its receptor C-X-C motif chemokine receptor 4 (CXCR4), fibronectin, and intercellular adhesion molecule 1 (ICAM-1) ligand [32]. While very late antigen-4 (VLA-4) and vascular cell adhesion molecule-1 (VCAM-1) binding promote cell arrest and tight adhesion, blocking either pathway leads to egress of plasma cells [29].

3.1.3 Bone marrow T lymphocytes

Mature T cells in the bone marrow are in constant exchange with the blood and make up a part of the total body recirculating lymphocyte pool that also includes the thoracic duct, spleen, and lymph nodes [28]. In the absence of ongoing immune responses or inflammations, homing to the bone marrow is a "default" pathway for the maintenance of recirculating memory T cells.

Regulation of bone marrow T-cell niches differs between steady-state conditions, immune response, and different disease states. Recirculating T lymphocytes compete with resident T cells for the same niche locations in the bone marrow [17]. The proliferation rates of cluster of differentiation 4 (CD4) and cluster of differentiation 8 (CD8) T cells are higher in the bone marrow than in the spleen and lymph nodes. T cells in the bone marrow have high expression of CXCR4, C-C chemokine receptor type 5 (CCR5), C-X-C motif chemokine receptor 6 (CXCR6), and C-X3-C motif chemokine receptor 1 (CX3CR1), but not of C-X-C motif chemokine receptor 3 (CXCR3). They respond to inflammatory chemokines, including C-C motif chemokine ligand 3 (CCL3), C-C motif chemokine ligand 4 (CCL4), C-C motif chemokine ligand 5 (CCL5), C-X-C motif chemokine ligand 16 (CXCL16), and C-X3-C motif chemokine ligand 1 (CX3CL1).

The recirculation of CD8 T cells involves rolling in bone marrow microvessels through L-, P-, and E-selectins. They stick to endothelial cells by the lymphocyte integrin α4β1, activated by SDF-1 (CXCL12) and the endothelial adhesion molecule VCAM-1. Modulating the expression of T-cell receptor CXCR4 by antigen, interleukin (IL)-2, and tumor necrosis factor (TNF) promotes the directed motility of T cells [17]. Granulocyte colony-stimulating factor (G-CSF) mobilizes regulatory T cells from the bone marrow.

3.2 T lymphocytes networking with other bone marrow elements

In the bone marrow, T lymphocytes contribute to the homeostasis of hematopoiesis and bone. T cells act mainly on the proliferative/differentiation phase of myelopoiesis rather than the stem cell maintenance and the bone remodeling system [18]. Many of the molecular signals in the immune response also contribute to bone homeostasis. Examples of osteoclast activators include interleukin-17 (IL-17) and receptor activator of nuclear factor κB ligand (RANK-L), expressed by activated CD4 and CD8 T cells. On the other hand, interferon-ɣ (IFN-ɣ) and interleukin-4 (IL-4) released by T cells inhibit bone resorption [17].

4. Dynamics of bone marrow lymphoid cells in health and disease

Bone marrow lymphocytes have central roles in many physiological and pathological processes. Memory B cells contribute to long-term antibody production, inflammations, tissue repair, and bone metabolism. T cells interact with other marrow elements and modulate their functions such as mesenchymal stromal cells, osteoclasts, osteoblasts, and hematopoietic precursors [17].

The lymphocyte traffic, which maintains homeostasis of the immune system, requires spatial and functional versatility. The cellular redistribution of mitochondria is an important factor in regulating these mechanisms permitting the motility of emigrating cells [33].

4.1 Aging

Aging alters basic cellular mechanisms leading to genomic instability, telomere shortening, epigenetic dysregulation, and cellular senescence, all contributing to a deranged hematopoiesis and immune system and increasing inflammatory diseases in the elderly. Common effects include reduced B cells and their subsets and impairment of antibody responses. The reduced generative capacity and altered microenvironment reduce HSC self-renewal and their preferential differentiation toward myeloid cells, B-1 and B-2 subsets, and regulatory B cells (Bregs), naive T cell, production with an increase in regulatory T cells (Tregs) [34].

Inflammatory changes in aging convert myeloid cells into pro-inflammatory 4-1BB ligand (4-1BBL)-expressing cells, leading to the activation of CD8+ T cells secreting antitumor granzyme B. The reduction of B-cell subsets suppresses antibody affinity and diversity and impairs antibody responses, with the expansion of age-related B cells (ABCs) contributing to inflammation via pro-inflammatory T-cell activation and cytokine release [34]. The downregulation of XBP-1 and B-lymphocyte-induced maturation protein-1 (Blimp-1) transcription factors, and upregulation of the plasma cell-inhibiting factor PAX-5, reduce B1 and spontaneously IgM-secreting B-1 cells and their diversity in the elderly.

On the other hand, the reduced E47 messenger RNA (mRNA) stability and activation-induced deoxycytidine deaminase (AID) transcription inhibit B-2 cell functions and isotype switching [34]. ABCs uniquely express cluster of differentiation 11b (CD11b), cluster of differentiation 11c (CD11c), and T-bet, and respond to innate activation stimuli, such as toll-like receptor 7 (TLR7) signals. They also secrete autoreactive antibodies and anti-inflammatory cytokine interleukin-10 (IL-10), reflecting their primary immunoregulatory function. They have distinct subsets that vary with extrinsic factors, such as age and antigen load. A major subset is a T-bet expressing TH1 cells that expand in response to innate stimuli, such as TLR7 and toll-like receptor 9 (TLR9) ligands. T-bet expression by B cells promotes immunoglobulin G2a (IgG2a) antibody isotype switching. A smaller subset is T-bet negative expressing C-X-C Motif Chemokine Receptor 5 (CXCR5) and it causes higher levels of immuno-globulin G1 (IgG1). Double-negative memory B cells may represent an intermediate stage from ABC to plasma cells. The age-related adipose-resident B cells (AABs) express pro-inflammatory markers [34].

4.2 Modulation of hematopoiesis in inflammatory conditions and infections

Significant alterations in leukocyte production occur in many inflammatory conditions, with an increase in granulopoiesis at the expense of lymphopoiesis. The mechanisms underlying such alterations are mostly due to modulation of the bone marrow microenvironment by stress signals and reduction of growth and retention factors, particularly stem cell factor and CXCL12, which preferentially inhibit lymphopoiesis [24]. However, overstressing hematopoietic stem and progenitor cells (HSPCs) in chronic inflammation leads to cell (DNA) damage and bone marrow failure. Furthermore, toxic insults to bone marrow stroma, e.g., chemotherapy may result in clonal hematopoiesis and potentially malignant transformation.

Figure 2 represents the inflammatory bone marrow microenvironment and its consequences [35]. An inflammatory microenvironment changes mesenchymal stem cells (MSCs) into an inflammatory, secretory phenotype, e.g., nestin$^+$, Gli1$^+$, and leptin-receptor+ (LepR) cells with the release of pro-inflammatory signals that alter the HSC niche and the erythroid precursors. Consequently, alterations in HSC function and output, including myeloid cell expansion, innate immune cells'/myeloid-derived suppressor cells' (MDSCs') recruitment, increased platelets' release by megakaryocytes and decreased the development of lymphoid cells. The impaired erythroid differentiation accounts partially for the associated anemia and the macrophage inflammatory phenotype further contributes to microenviron-ment inflammation. Adipocytes also increase with the release of inflammatory signals [35].

Bone marrow T cells can also modulate hematopoiesis in inflammatory condi-tions. At low neutrophil counts, interleukin 23 (IL-23) released by macrophages and DCs stimulates CD4 T cells, gamma delta T cells (γδ-T cells), and NK cells, produc-ing IL-17, which promotes granulopoiesis. At high neutrophil counts, inhibition of interleukin-23 (IL-23) by apoptotic neutrophils results in negative feedback [17]. Sometimes, the bone marrow is the target of effector T cells, such as in hematologi-cal malignancies, idiopathic thrombocytopenic purpura (ITP), and autoimmune diseases [17].

The brief neutropenia occurs early in sepsis and rapidly corrects by accelerated cell emigration and reactive neutrophilia follows mainly due to increased immature proliferative compartment of granulopoiesis. Immature neutrophils in the bone mar-row also become resistant to inflammatory signals [24]. An early onset lymphopenia, characteristic of sepsis, is mainly due to massive lymphocyte apoptosis primarily of CD4+ T cells by activation of multiple cell death pathways [18]. Inflammatory cytokines also promote lymphocyte mobilization [24]. Homing and proliferation of CD4+ memory T cells occur in response to IL-7 at resolution of sepsis-induced lymphopenia.

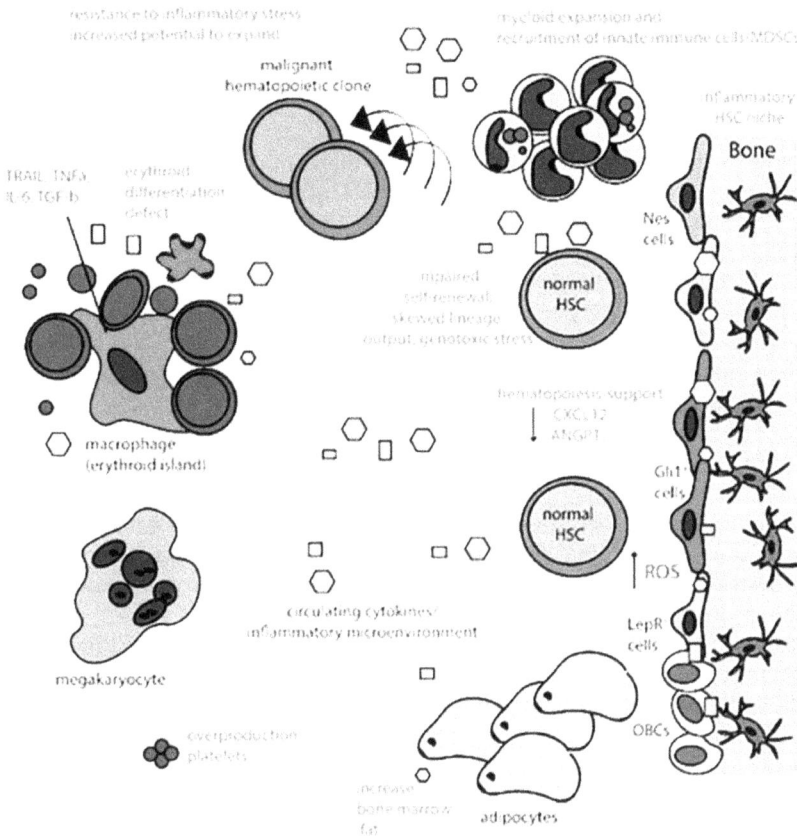

Figure 2.
The inflammatory bone marrow microenvironment. An inflammatory bone marrow microenvironment develops when MSCs acquire an inflammatory, secretory phenotype such as nestin+ (Nes+), Gli1+, leptin-receptor (LepR+) cells and release pro-inflammatory signals affecting both the HSC niche and the erythroblastic islands [35].

4.3 Leukocyte dynamics in stroke and stress conditions

Stroke and stress often induce lymphopenia and immunosuppression while increasing myeloid cells.

Lymphopenia affects the entire B-cell lineage, including B-cell progenitors in the bone marrow, due to reduced proliferation and differentiation, and enhanced apoptosis. Lymphopenia in these cases is due to the hypothalamic-pituitary-adrenal (HPA) axis action via glucocorticoid receptors on hematopoietic cells. Both long-range and local microenvironmental signals link the neuroendocrine system to emergency hematopoiesis and impair the lymphopoiesis. The neutrophil-to-lymphocyte ratio in peripheral blood is a prognostic indicator that defines poor long-term outcomes in these patients [36].

Cell death and release of danger-associated factors, such as high mobility group box 1 (HMGB-1) and heat shock proteins (HSPs) in ischemic injuries, provoke an inflammatory response. In addition, systemic glucocorticoid levels caused by activation of the sympathetic nervous system (SNS) and the HPA axis increase the apoptosis of B-cell progenitors [37]. Furthermore, altered lineage decision of HSCs in ischemia is due to sympathetic signals to stem cell niches, favoring myelopoiesis.

4.4 Hematopoietic stem cell engraftment

In steady states, lymphocyte homeostasis is maintained by balancing production and lifespan rates. A full recovery of T cells is often slow after autologous HSC transplantation and may take several years [38].

T cells can promote HSC engraftment by eliminating residual host cells responsible for rejection and enhancing stromal function [17].

Although studying the dynamics of transplanted cells is often difficult, recent imaging techniques allow *in vivo* tracking of their migration and differentiation, including "optical coherence tomography," "two photon-exited fluorescence microscopy," and "confocal microscopy" [39].

4.5 Bone metastasis

The complex network of immune and bone cells of the bone marrow microenvironment increases the chance of metastases to bone than other sites [40]. The endosteal and vascular stem cell niches of the marrow microenvironment support the seeding and establishment of cancer cells [41].

The tumor antigen-primed T cells migrate to the bone marrow and initiate the pre-metastatic niche by the transfer of signals that alter bone homeostasis. The tumor cells then colonize the bone marrow and reside in the hematopoietic niches [19].

The interaction of tumor cells with bone is the main pathogenic mechanism of metastasis involving released signals that sustain a vicious cycle [42]. In addition, the release of immune-suppressive cytokines by tumor cells promotes the conversion of M1 macrophages and N1 neutrophils to tumor-associated M2 macrophages and N2 neutrophils, with tumor-promoting activity [40].

Another mechanism-promoting metastasis is immunosuppression, in which myeloid-derived suppressor cells (MDSCs) play a critical role. They are immature myeloid cells co-expressing CD11b and cluster of differentiation 33 (CD33) [43], originating in the bone marrow and migrating to secondary lymphoid organs, where they inhibit the CD8+ T antitumor response [41]. MDSCs also suppress T-cell function by impairing antigen recognition, releasing small soluble oxidizers, and depleting local essential amino acids. They suppress both CD4 and CD8 T cells while promoting activation and expansion of regulatory T cells (Tregs) [44]. The critical roles of MDSCs rationally explain their assessment as prognostic indicators in osteotropic tumors.

Meanwhile, cancer cell-released soluble factors stimulate MDSCs differentiation into activated osteoclasts. Both MDSCs and activated T cells produce pro-osteoclastogenic factors, including C-C chemokine receptor type 2 (CCR2) and RANK-L, respectively. Thus, tumor growth sustains the osteoclast activation by several mechanisms [41].

In summation, tumor-cell intrinsic traits and interaction with the specialized microenvironmental niches play critical roles in controlling tumor-cell colonization with initial seeding, dormancy, and outgrowth, as demonstrated in **Figure 3** [45]. In the perivascular niche, tumor cells interact with CXCL12-expressing stromal cells and endothelial E-selectin promotes "mesenchymal-to-epithelial transition," "stemness," "survival," and "growth." Activated osteoclasts express Notch ligand, Jagged1 (JAG1),

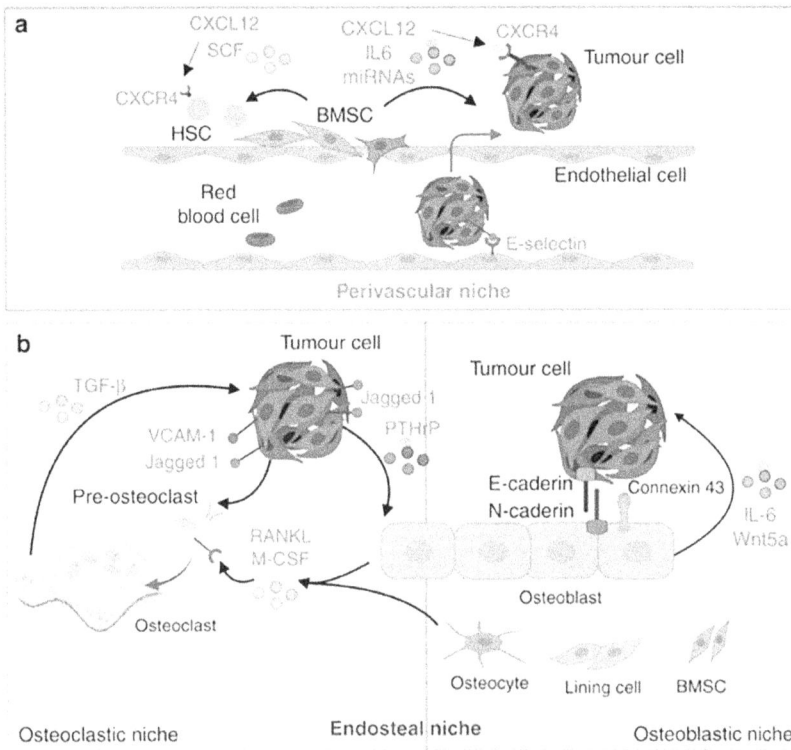

Figure 3.
Metastatic bone marrow niches. The perivascular and endosteal niches in bone metastasis [45].

and VCAM-1, increasing the production of the osteoclast-stimulating factors, macrophage colony-stimulating factor (M-CSF), and receptor activator of nuclear factor κB ligand (RANK-L) which alter the bone turnover. In addition, bone resorption releases transforming growth factor-β (TGF-β) propagating a "vicious cycle" and promoting osteolytic bone metastasis.

On the other hand, osteoblasts secrete Wnt5a and interleukin (IL)-6 and form gap junctions, and E-cadherin/N-cadherin junctions, thus promoting bone metastasis [45].

Other cells contributing to the metastatic niche include megakaryocytes, adipocytes, and sympathetic nerve cells by altering the immune escape, dormancy, and proliferation of tumor cells . Understanding the intricate pathways of metastasis helps the evolution of new therapeutic targets [45].

5. Conclusion

In conclusion, the unique bone marrow's organizational and cellular structure is critical to the immune system's homeostasis and functions. Understanding the lymphocyte dynamics in the bone marrow and their regulations may clarify many of the pathologic changes in many physiologic and pathologic conditions and their prognostic impact. It can also provide potential venues for the prevention and treatment of some diseases.

Author details

Samia Hassan Rizk
Laboratory Hematology, Clinical Pathology Department, Cairo University School of
Medicine, Egypt

*Address all correspondence to: rizksh@gmail.com

IntechOpen

References

[1] Immunopaedia. "Free Immunology Education." 2022. Available from: www.immunopaedia.org.za/

[2] Moreau JM. Chapter 6. In: B Cell Dynamics within the Bone Marrow Microenvironment: Comparisons of Inflammation and Steady State [Thesis]. Canada. TSpace: Department of Immunology, University of Toronto; 2017. pp. 68-100. Available from: https://tspace.library.utoronto.ca/handle/1807/77479

[3] Christopher MJJ, Rao Liu F, Woloszynek JR, Link DC. Expression of the G-CSF receptor in monocytic cells is sufficient to mediate hematopoietic progenitor mobilization by G-CSF in mice. The Journal of Experimental Medicine. 2011;**208**(2):251-260. DOI: 10.1084/jem.20101700

[4] Kawano Y, Petkau G, Wolf I, Tornack J, Melchers F. IL-7 and immobilized, kit-ligand stimulate serum and stromal cell-free cultures of precursor B-cell lines and clones. European Journal of Immunology. 2017;**47**:206-212. DOI: 10.1002/eji.201646677. PMID: 17145957; PMCID: PMC2118173

[5] Pinho S, Tony MT, Yang E, Wei Q, Claus NC, Frenette PS, et al. Lineage-biased, hematopoietic stem cells are regulated by distinct niches. Developmental Cell. 2018;**44**:634-641, Cell Press. Elsevier Inc. DOI: 10.1016/j.devcel.2018.01.016

[6] Lucas D. Structural organization of the bone marrow and its role in hematopoiesis. Current Opinion in Hematology. 2021;**28**(1):36-42. DOI: 10.1097/MOH.0000000000000621

[7] Grosschedl R. Establishment and maintenance of B cell identity. In: Cold Spring Harbor Symposia on Qualitative Biology. Vol. LXXVIII, 78. Cold Spring Harbor Laboratory Press; 2013. pp. 23-30. DOI: 10.1101/sqb.2013.78.020057. Available from: https://www.researchgate.net/publication/261736570

[8] Kaminski N, Sulentic C. B lymphocytes. In: Encyclopedic Reference of Immunotoxicology. Berlin, Heidelberg: Springer; 2005. pp. 88-89. DOI: 10.1007/3-540-27806-0_158

[9] Salvo P, Vivaldi FM, Bonini A, Biagini D, Bellagambi FG, Miliani FM, et al. Biosensors for detecting lymphocytes and immunoglobulins. Biosensors. 2020;**10**:155. DOI: 10.3390/bios10110155 https://pubmed.ncbi.nlm.nih.gov/33946495/

[10] Cherukommu S. Role of GSK-3 and T-Bet in Anti-Tumor Immunity [Thesis]. Canada: Université de Montréal; 1980]. Faculté de Médecine Mémoire présenté en vue de l'obtention du grade de Maitrise en biologie moléculaire. Papyrus: Institutional Repository. 2021. Available from: https://papyrus.bib.umontreal.ca/xmlui/handle/1866/25645

[11] Verstegen NJM, Pollastro S, Unger PA, Marsman C, Elias G, Jorritsma T, et al. Single-cell analysis reveals dynamics of human B cell differentiation and identifies novel B and antibody-secreting cell intermediates. Elife. 2 Mar 2023;**12**:e83578. DOI: 10.7554/eLife.83578. PMID: 36861964; PMCID: PMC10005767

[12] Mehr R, Shahaf G, Sah A, Cancro M. Asynchronous differentiation models explain bone marrow labeling kinetics and predict reflux between the pre-and immature B cell pools. International Immunology. 2003;**15**(3):301-312.

The Japanese Society for Immunology. DOI: 10.1093/intima/dxg025 Available from: www.intimm.oupjournals.org

[13] Zehentmeier S, Roth K, Cseresnyes Z, et al. Static and dynamic components synergize to form a stable survival niche for bone marrow plasma cells. European Journal of Immunology. 2014;**44**:2306-2317. DOI: 10.1002/eji.201344313

[14] Winter O, Moser K, Mohr E, et al. Megakaryocytes constitute a functional component of a plasma cell niche in the bone marrow. Blood. 2010;**116**(11):1867-1875, ISSN 0006-4971. DOI: 10.1182/blood-2009-12-259457 Available from: https://www.sciencedirect.com/science/article/pii/S000649712032989X

[15] Belnoue E, Tougne C, Rochat AF, et al. Homing and adhesion patterns determine the cellular composition of the bone marrow plasma cell niche. Journal of Immunology (Baltimore, Md.: 1950). 2012;**188**(3):1283-1291. DOI: 10.4049/jimmunol.1103169

[16] Caraux A, Vincent L, Bouhya S, et al. Residual malignant and normal plasma cells shortly after high dose melphalan and stem cell transplantation. Highlight of a putative therapeutic window in multiple myeloma? Oncotarget. 2012;**3**(11):1335-1347. DOI: 10.18632/oncotarget.650

[17] Di Rosa F. T-lymphocyte interaction with stromal, bone, and hematopoietic cells in the bone marrow. Immunology and Cell Biology. 2009;**87**:20-29. DOI: 10.1038/icb. Available from: http://www.nature.com/icb/journal/v87/n1/full/icb200884a.html

[18] Skirecki T, Swacha P, Hoser G, et al. Bone marrow is the preferred site of memory CD4 + T cell proliferation during recovery from sepsis. JCI Insight.

2020;**5**(10):e134475. DOI: 10.1172/jci.insight.134475.insight.jci.org

[19] Bonomo A, Monteiro AC, Balduíno A. Hematopoietic Stem Cells, Tumor Cells, and Lymphocytes — Party in the Bone Marrow. London, UK: InTechOpen; 2014. DOI: 10.5772/58843

[20] Bilwani FA, Knight KL. Adipocyte-derived soluble factor(s) inhibits early stages of B lymphopoiesis. London, UK: Journal of Immunology. InTechOpen; 2014;**189**(9):4379-4386. DOI: 10.4049/jimmunol.1201176 Epub 2012 Sep 21

[21] Kim SE, Kim HM, Doh J. Single-cell arrays of hematological cancer cells for assessment of lymphocyte cytotoxicity dynamics, serial killing, and extracellular molecules. Lab on a Chip. 2019;**19**:2009-2018. DOI: 10.1039/c91c00133f

[22] Tsunokuma N, Yamane T, Matsumoto C, et al. Depletion of neural crest-derived cells leads to reduction in plasma noradrenaline and alters B Lymphopoiesis. Journal of Immunology. 2016;**198**(1):156-169. DOI: 10.4049/jimmunol.1502592

[23] Dudakov JA, Goldberg GL, Reiseger JJ, et al. Sex steroid ablation enhances hematopoietic recovery following cytotoxic antineoplastic therapy in aged mice. Journal of Immunology. 2009;**183**:7084-7094

[24] Ueda Y, Kondo M, Kelsoe G. Inflammation and the reciprocal production of granulocytes and lymphocytes in bone marrow. Journal of Experimental Medicine. The Rockefeller University Press; 6 Jun 2005;**201**(11):1771-1780. DOI: 10.1084/jem.20041419. PMID: 15939792; PMCID: PMC1952536

[25] Pabst R. The bone marrow is not only a primary lymphoid organ: The critical

role for T lymphocyte migration and housing of long-term memory plasma cells. European Journal of Immunology. 2018;**48**:1096-1100. DOI: 10.1002/eji.201747392 (2018)

[26] Matthes-Martin S, Feuchtinger T, Shaw PJ, et al. European guidelines for diagnosis and treatment of adenovirus infection in leukemia and stem cell transplantation: Summary of ECIL-4. Transplant Infectious Disease. 2012;**14**:555-563

[27] Tikhonova AN, Dolgalev I, Hu H, et al. The bone marrow microenvironment at single-cell resolution. Nature. 2019;**569**(7755):222-228. DOI: 10.1038/s41586-019-1104-8 [Epub 2019, Apr 10]. Erratum in: Nature. 2019 Aug;572(7767): E6

[28] Schürch CM, Caraccio C, Nolte MA. Diversity, localization, and (patho) physiology of mature lymphocyte populations in the bone marrow. Blood. 2021;**137**(22):3015-3026. DOI: 10.1182/blood.2020007592

[29] Paramithiotis E, Cooper MD. Memory B lymphocytes migrate to bone marrow in humans. Proceedings of the National Academy of Sciences of the United States of America. 1997;**94**:208-212

[30] Gatter K, Brown D. Bone Marrow Diagnosis: An Illustrated Guide. 3rd ed. Chichester, UK: John Wiley & Sons Ltd; 2014. ISBN 10: 1118253655; ISBN 13: 9781118253656

[31] Fauci AS. Human bone marrow lymphocytes I. Distribution of lymphocyte subpopulations in the bone marrow of normal individuals. The Journal of Clinical Investigation. 1975;**56**:98-110

[32] Benet Z, Jing Z, David R. Fooksman DR. Plasma cell dynamics in the bone marrow niche. 2021, Cell Reports 34, 108733. DOI:10.1016/j.celrep.2021.108733

[33] Campello S, Lacalle RA, Bettella M, et al. Orchestration of lymphocyte chemotaxis by mitochondrial dynamics. Journal of Experimental Medicine. Vol. 203, No. 13. The Rockefeller University Press; 25 Dec 2006. pp. 2879-2886. DOI: 10.1084/jem.20061877. Epub 2006 Dec

[34] De Mol J, Kuiper J, Tsiantoulas D, Foks AC. The dynamics of B cell aging in health and disease. Frontiers in Immunology. 2021;**12**:733566. DOI: 10.3389/fimmu.2021.733566

[35] Leimkühler NB, Schneider RK. Inflammatory bone marrow microenvironment. Hematology. American Society of Hematology. Education Program. 2019;**2019**(1):294-302. DOI: 10.1182/hematology.2019000045

[36] Courties G, Frodermann V, Lisa Honold L, et al. Glucocorticoids regulate bone marrow B Lymphopoiesis after stroke. Circulation Research. 2019;**124**:1372-1385. DOI: 10.1161/CIRCRESAHA.118.314518

[37] Dai S, Mo Y, Wang Y, Xiang B, Liao Q, Zhou M, et al. Frontiers in Oncology. 2020;**10**:1492. DOI: 10.3389/fonc.2020.01492

[38] Baliu-Pique M, van Hoeven V, Drylewicz J, et al. Cell-density, independent increased lymphocyte production and loss rates post-autologous HSCT. eLife. 2021;**10**:e59775. DOI: 10.7554/eLife.59775

[39] Ahn S, Choe K, Lee S, Kim K, Song E, Seo H, et al. Intravital longitudinal wide-area imaging of dynamic bone marrow engraftment and multilineage differentiation through

nuclear-cytoplasmic labeling. PLoS One. 2017;**12**(11):e0187660. DOI: 10.1371/journal.pone.0187660

[40] Xiang L, Gilkes DM. The contribution of the immune system in bone metastasis pathogenesis. International Journal of Molecular Sciences. 2019;**20**(4):999. DOI: 10.3390/ijms20040999

[41] D'Amico L, Roato I. The impact of immune system in regulating bone metastasis formation by Osteotropic tumors. Journal of Immunology Research. 2015;**2015**:143526. DOI: 10.1155/2015/143526

[42] Gadiyar V, Patel G, Chen J, Vigil D, Ji N, Campbell V, et al. Targeted degradation of MERTK and other TAM receptor paralogs by heterobifunctional targeted protein degraders. Frontiers in Immunology. 2023;**14**:1135373. DOI: 10.3389/fimmu.2023.1135373

[43] Georgoulis V, Papoudou-Bai A, Makis A, Kanavaros P, Hatzimichael E. Unraveling the immune microenvironment in classic Hodgkin lymphoma: Prognostic and therapeutic implications. Biology. 2023;**12**:862. DOI: 10.3390/biology12060862

[44] Satolli MA, Buffoni L, Spadi R, Roato I. Gastric cancer: The times, they area-changin. World Journal of Gastrointestinal Oncology. 2015;7(11):303-316. ISSN: 1948-5204 (online). DOI: 10.4251/wjgo.v7.i11.303. Available from: http://www.wjgnet.com/1948-5204/full/v7/i11/303.htm

[45] Chen F, Han Y, Kang Y. Bone marrow niches in the regulation of bone metastasis. British Journal of Cancer. 2021;**124**:1912-1920. DOI: 10.1038/s41416-021-01329-6

Chapter 3

Head and Neck Lymphadenopathy in Oral Cancer

Ankita Tandon, Kumari Sandhya and Narendra Nath Singh

Abstract

Ranging from localised to generalised, infectious to neoplastic, autoimmune, or miscellaneous aetiology; lymphadenopathies have a wide array of clinical presentations. Assessment of the true pathobiology of lymphadenopathies is a challenging process specially cases with lymphadenopathy due to malignancies in the head and neck region. A multitude of masking signs and symptoms make it even more complicated. However, a correct diagnostic workflow facilitates easy evaluation of such lymphadenopathies. Although, the correct clinical examination may help to achieve correct diagnosis in some lymphadenopathy cases, some suspicious and unexplained lymphadenopathies warrant further investigations. This chapter clearly focuses on the clinical, diagnostic, and histopathologic spectrum of head and neck lymphadenopathies arising in oral cancer and stressing upon the pathways of lymphatic spread of malignancy along with a multitude of lymph node characteristics which play a key role in diagnosis.

Keywords: cervical, head and neck, lymphadenopathy, lymph nodes, oral squamous cell carcinoma

1. Introduction

Oral Squamous Cell Carcinoma (OSCC) is the sixth most commonly diagnosed cancer among all cancer types [1, 2]. While smoking, drinking, and HPV infection are known risk factors, genetics also plays a significant influence in tumour development, progression, and patient's response to therapy [1, 3].

Lymphatic capillaries, afferent lymphatic vessels, lymph nodes, efferent lymphatic vessels, and diverse lymphoid organs are only a few of the anatomical parts that make up the lymphatic system. Lymph nodes (LNs) are tiny bean-shaped structures that line lymphatic channels. They act as a filter, check lymphatic fluid/blood composition, drain extra tissue fluid and plasma protein leaks, absorb pathogens, boost immune response, and get rid of infection [4]. As OSCC develops, it frequently metastasizes to nearby cervical LNs. During variable clinical assessment, it has been observed that 30 to 50 percent of OSCC patients had metastatic LN [5]. For proliferation, invasiveness, and metastasis in solid tumours, ECM degradation and reconstruction in the stroma and adjacent tissues are essential. Matrix metalloproteinases (MMPs) and their regulators, which come from tumour cells and stromal cells such fibroblasts, macrophages, dendritic cells, and neutrophils, cause the ECM to remodel. Lymphatic metastasis involves the expression of multiple MMPs and the variables that are connected with them for metastasis and prognosis [5].

Cervical lymph node metastasis has always been a harbinger of poor prognosis, with node-positive status immediately upstaging disease to stage 3 or higher and halving 5-year survival rates [6]. It is widely acknowledged that cervical lymph node involvement affects prognosis for HNSCC, and even one positive lymph node can result in a 50% decrease in overall survival [7]. Additionally, having affected lymph nodes has significant short-term therapy effects. For long-term survival and recurrence-free survival in patients with OSCC, the number of lymph nodes involved, the anatomical levels of positive lymph nodes, the size of the metastases, the presence of microscopic or macroscopic extracapsular spread, and soft tissue deposits are all significant prognostic factors [8].

This chapter therefore intends to focus upon the key determinants to head and neck lymphadenopathy in oral cancer to upscale the current evidence for its usage in determining the best therapeutic overlay for OSCC patients.

2. What are lymph nodes and the lymphatic system?

The kidney-shaped organs known as lymph nodes are positioned in groups all throughout the body, mainly in the groin, armpits, neck, and centre of the chest and belly. Lymphatic canals link lymph nodes to one another. Through lymphatic channels, lymphatic fluid travels from all bodily tissues to neighbouring lymph nodes, which act as a kind of filter. The immune cells in the lymphatic system known as lymphocytes can proliferate when the immune system is engaged, such as with infections or cancer. This results in lymphadenopathy, which is the enlargement of one or more lymph nodes [9].

3. What is lymphadenopathy?

Lymphadenopathy refers to nodes that are abnormal in either size, consistency or number. The global incidence of patients reporting unexplained lymphadenopathy ranges between 75% of localised to 25% of generalised types. It may be classified as: (1) localised when lymph nodes of only one area is involved (55% are exclusive to Head and neck region); and (2) generalised if lymph nodes are enlarged in two or more noncontiguous areas [10].

4. Sentinel node

The earliest lymph node(s) to drain a primary tumour are called sentinel lymph nodes [11]. It is still debatable whether sentinel lymph node biopsy (SLNB) or elective neck dissection (END) should be used to stage patients with early OSCC (T1 and T2 N0 disease) [12]. In the event that a lymphogenic tumour spreads, the sentinel lymph nodes (SLNs), which must be located and removed, are more likely to contain metastases. If metastatic tumour deposits are discovered in the SLN, further treatment (surgery and/or radiotherapy) of the nodal basin should be performed because the SLN's histological condition reflects that of the rest of the nodal basin [13].

There is rising evidence in the literature that supports obtaining more sections or employing immunohistochemistry to detect micro metastases can improve the diagnosis of nodal metastases. The sensitivity of metastasis identification is increased by combining various methods [8].

5. Tumour biology as related to metastasis

A variety of genotypic, phenotypic and microenvironmental factors conspire to define the metastatic potential of a tumour. Previously, it was believed that cancer cells entered the lymphatic system through pre-existing lymphatic capillaries close to the tumour (**Figure 1a**). Solid tumours can stimulate lymphangiogenesis, according to recent research on animal models (**Figure 1b**). In this situation, lymph node metastasis has been linked to intra-tumoral and peritumoral lymphangiogenesis. By binding to VEGFR3, a tyrosine kinase receptor expressed on the surface of lymphatic endothelial cells, VEGF-C/D released by the tumour has been demonstrated to play a significant role in lymphangiogenesis. The existence and biological function of lymphatics within experimental and human tumours have remained debatable despite mounting evidence indicating an active role of VEGF-C- or VEGF-D-induced tumour lymphangiogenesis in cancer spread to local lymph nodes. According to the lack of lymphatic uptake of tracers that were injected near experimental tumours, high interstitial pressure within tumours has been postulated to inhibit intra-tumoral lymphatic vessel formation and function [14].

6. Mechanism of lymphatic metastasis

The majority of research supported the idea that lymphangiogenesis generated by tumours and greater levels of lymphangiogenic growth factors are linked to increased

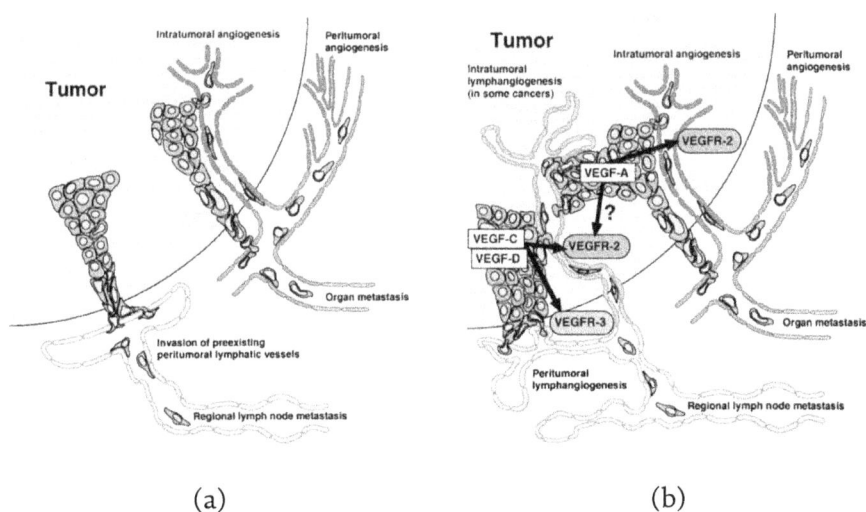

(a)   (b)

Figure 1.
Model of tumour metastasis: (a) traditional model, and (b) active model [14].

rates of LN metastasis and a poor prognosis. PDGF-BB, IGF1 and -2, FGF2, HGF, angiopoietin-2, sphingosine-1-phosphate, adrenomedullin, and IL-7 are among the additional mediators that have been associated to the development of lymphangiogenesis in cancers and other disorders in addition to growth factors from the VEGF family. Further investigation is still needed to determine the relative importance of these factors in contrast to VEGFs for various cancer types. The VEGF-C, VEGF-D, VEGF-A, and HGF produced by the main tumour are picked up by peritumoral lymphatic capillaries and delivered to the SLNs via collecting lymphatics, where they operate directly on preexisting lymphatic vessels in a manner similar to inflammation-induced lymphangiogenesis (**Figure 2A** and **B**).

The remodelling and SMC rearrangement of distant (post-SLN) lymphatic vessels and LNs, as well as secondary metastasis, such as organ metastasis, have recently been observed. Metastatic tumour cells represent a significant source of lymphangiogenic factors, such as VEGF-C, once they have spread to their draining LNs [11].

Cancer cells with characteristics similar to stem cells might also find a home in lymphatic endothelium (**Figure 2C**). Lymphatic endothelium may offer a protective milieu for long-term tumour cell survival, according to clinical observations of so-called "in-transit metastases," or metastatic tumours that form in lymphatic arteries between the source tumour and the draining LN. Moreover, when the original tumour has been removed, tumour cells may persist in dormancy within draining LNs for a long time [11, 15].

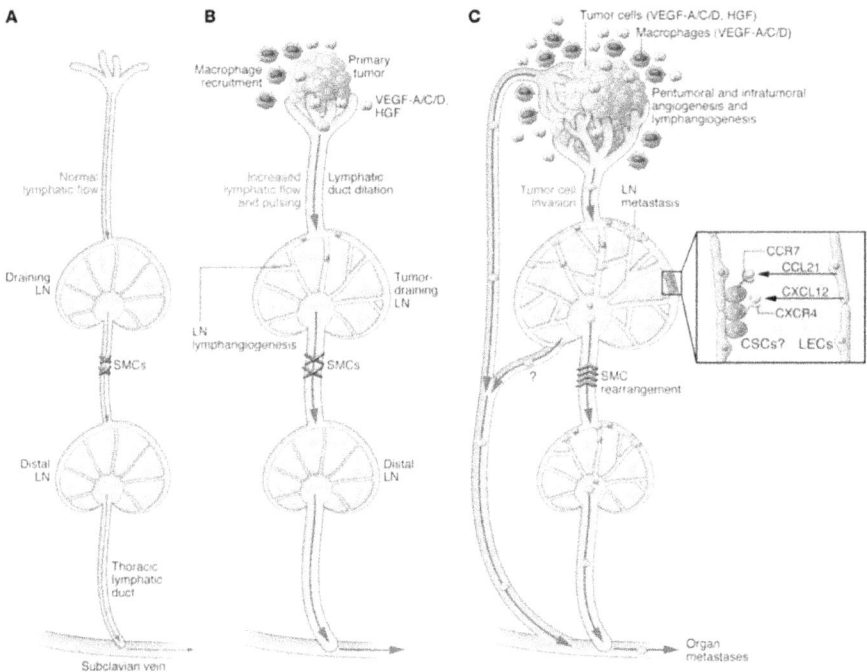

Figure 2.
An important contribution of tumour and LN lymphangiogenesis to cancer metastasis. (A) Normal lymphatic tissue drainage through lymphatic capillaries, collecting lymphatics, and LNs, (B) lymphangiogenic factors produced by premetastatic tumours, and (C) once metastatic tumour cells have spread to their draining LNs, they serve as a major source of lymphangiogenic factors which promote secondary metastasis, including organ metastasis [11].

7. Head and neck cancer metastatic spread and lymphatic drainage patterns

The major lymphatic pathway, the posterior pathway, the anterior lymphatic pathway, and the superficial-lateral pathway are the four functional drainage pathways for the cervical lymphatics [16]. After tumour excision, lymph node metastasis greatly raises the risk of systemic cancer spread and recurrence and reduces the effectiveness of treatment, especially when extranodal extension is evident [17, 18]. As a result, the accurate staging and diagnosis of nodal metastases play a crucial role in determining the prognosis and treatment of head and neck cancer. In the N0 neck, for example, sampling of the sentinel node can identify patients who actually require a neck dissection while sparing others who do not show signs of disease [19].

8. The pathophysiology of lymphatic metastasis by head-and-neck cancers

In order to continue invading the surrounding tissue during metastasis, cancer cells must first penetrate the basement membrane of the epithelium from where they originated. Tumour cells may potentially infiltrate and move through lymphatic channels after invading surrounding lymphatics and metastasis to lymph nodes [20]. The lymph fluid carries the tumour cells to the closest lymph node when they reach the lymphatic system. The afferent lymphatic vessels guide these cells into the subcapsular sinus, which is the region immediately below the lymph node capsule where lymphocytes and antigen-presenting cells circulate [21]. The initial site of metastasis may take place here, in this sinus [22]. Tumour cells may create new colonies farther along the nodal chain as the metastatic colony expands. As nodal metastases expand and invade, the nodal endothelium may also be damaged, leading to extracapsular tumour invasion [22]. The latter symptoms are indicative of an advanced stage of metastatic dissemination and, for the majority of human malignancies, a dismal prognosis. While hematogenous disease spread is less frequent, head and neck squamous cell carcinoma (HNSCC), which accounts for 80–90% of tumours developing in the UADT of the head and neck [23], has a significant propensity for regional lymphatic metastasis. It's not quite obvious why lymphatic metastasis tends to be biased. According to one explanation, HNSCC prefers to produce early lymphatic metastases because its lymphatic system is so rich in compared to the rest of the body in the head and neck [24]. Metastatic tumour cells may enter lymphatic vessels more easily than into other nearby venous, arterial, or capillary vessels due to the absence of tight connections between lymphatic endothelial cells. Additionally, the lymph's gentle, low shearing flow makes the environment in which a travelling tumour cell or cell cluster lives less harsh, which greatly boosts the spread's effectiveness while lessening the difficulty of establishing new tumour-genic colonies [24]. Along with lymph node invasion, head and neck tumours may exhibit blood vessel invasion; however, this occurrence often results from extracapsular lymph node metastases that have subsequently invaded the vascular system. Of course, these findings are related to a more advanced condition [19, 25].

9. Skip metastases

Skip metastases are a discontinuous spread of cancer in which foci of involvement are interspersed among unaffected, adjacent regions. It majorly happens because of the rich lymphatic network of the majority of intraoral anatomic sites. Incidence of skip metastasis in OSCC to level IV and V bypassing levels I to II, ranges from 0.2 to 4.8% [26].

10. Occult nodal disease

Occult nodal disease represent metastatic deposits which are small enough to evade detection by clinical and radiographic methods and are found on microscopic analysis.

Such small tumour deposits require keen observation of all lymph nodes along with confirmation using specific IHC markers. In this reference, micro metastases are tumour deposits between 0.2–2 mm within lymph nodes whereas, Isolated Tumour Cells (ITCs) refer to single/small clusters of tumour cells '0.2 mm inside the lymph nodes [27]. It has been reported that micro metastasis occurs in 3–7% of nodes and 9–22% of patients with clinically N_0 necks and can be detected by molecular markers such as cytokeratin [28, 29]. Some strong predictors of occult nodes within the primary tumour may be desmoplasia with or without perineural infiltration, pT4a clinical stage and thickness of the tumour (≥ 4 mm). Tumour burden within the lymph nodes have always stood the test of time as a valuable prognostic factor [27].

11. Lymph node characteristics

11.1 Lymph node ratio

Recent research has looked at the lymph node ratio (LNR), sometimes also denoted as lymph node density, as a novel feature for assessing prognosis in patients with pN1 illness. More is the ratio, poorer is the prognosis [7]. LNR has become a stand-alone prognostic factor for a number of tumour forms, including HNSCC, as inadequate LN retrieval may cause pathological understaging. This ratio makes an effort to account for any probable prejudice in the sampling technique. It takes into account three variables that may have an overall impact on the staging of LNs: (1) tumour aspects, which represent the disease's actual regional spread (i.e., the number of lymph nodes that were actually positive); (2) surgical aspects (i.e., the number of nodes actually removed during neck dissection); and (3) detection aspects (i.e., the precision of the pathological analysis). Therefore, innovations like the greater use of immunohistochemistry, molecular methods, serial sectioning, and/or the discovery of tiny metastatic deposits in lymph nodes, as well as the tendency towards more limited/selective neck dissection, may also influence LNR [7, 30].

LNR is defined as the number of positive lymph nodes divided by the total number of lymph nodes excised, regardless of the extent of neck dissection [31]. LNR is a proxy for the sufficiency of neck dissection and corresponds majorly with total lymph node yield (LNY).

LNY (also referred to as lymph node harvest or lymph node count) is defined as the count of lymph nodes retrieved after neck dissection [32]. Numerous authors have discussed the importance of LNY and LNR as prognostic indicators, with higher LNY and lower LNR resulting in better survival [33]. Patients with lower LNRs who were node positive in several studies even had similar results to patients who were node negative. For instance, when compared to LNR and AJCC N staging, the number of positive nodes of more than five has greater prognostic influence [34].

11.2 Extranodal extension (ENE)

If the lymph node's metastatic tumour penetrates the lymph node capsule and into the nearby connective tissue, whether or not there is a stromal reaction present, the ENE is deemed positive. The maximum millimetre distance, in either intact or reconstructed nodal capsules, between the most distant point of invasion into extranodal tissue and the extent of ENE can be calculated [33]. Since ENE was added to the most recent version of the American Joint Committee on Cancer (AJCC) handbook, it is advised to categorise the extent of ENE in most head and neck cancers (HNC) as minor (ENEmi 2 mm from the capsule) and significant (ENEma >2 mm) [35, 36].

11.3 Level of lymph node involvement

Another lymph node characteristic that has been linked to a poor prognosis in numerous studies is the incidence of lymph node metastases in the lower levels of the neck i.e., upto levels IV and V [33].

11.4 Size of the tumour deposit within the lymph node

The relationship between the size of the tumour deposit and ENE has been documented in numerous investigations. According to some research, a deposit size of 11.5 mm would accurately and specifically indicate ENE. Additionally, there is a strong correlation between a tumour deposit that is less than 14 mm in size and reduced disease-free and overall survival [33].

11.5 Metastatic lymph node clearance (MLNC)

Because the total number of removed LN varies depending on the method of neck dissection, it is important to take this into account when calculating the Lymph Node Ratio (LNR) and Lymph Node Yield (LNY) in the clinical environment [37].

12. Conclusion

Lymphadenopathy in head and neck cancer present with a plethora of characteristics with significant implications on patient outcomes. The pathways of lymph node metastasis are distinct and represent a variety of histopathological interpretations having a significant bearing on patient prognosis.

Author details

Ankita Tandon[1]*, Kumari Sandhya[2] and Narendra Nath Singh[1]

1 Department of Oral Pathology, Microbiology, and Forensic Odontology, Dental Institute, RIMS, Ranchi, Jharkhand, India

2 Department of Anatomy, RIMS, Ranchi, Jharkhand, India

*Address all correspondence to: drankitatandon7@gmail.com

IntechOpen

References

[1] Mirza AH, Thomas G, Ottensmeier CH, King EV. Importance of the immune system in head and neck cancer. Head & Neck. 2019;**41**(8):2789-2800. DOI: 10.1002/hed.25716. Epub 2019 Feb 28

[2] Chaturvedi AK, Anderson WF, Lortet-Tieulent J, Curado MP, Ferlay J, Franceschi S, et al. Worldwide trends in incidence rates for oral cavity and oropharyngeal cancers. Journal of Clinical Oncology. 2013;**31**(36):4550-4559. DOI: 10.1200/JCO.2013.50.3870. Epub 2013 Nov 18

[3] Birkeland AC, Uhlmann WR, Brenner JC, Shuman AG. Getting personal: Head and neck cancer management in the era of genomic medicine. Head & Neck. 2016;**38**(Suppl. 1):E2250-E2258. DOI: 10.1002/hed.24132. Epub 2015 Aug 13

[4] Null M, Arbor TC, Agarwal M. Anatomy, lymphatic system. In: StatPearls. Treasure Island (FL): StatPearls Publishing; 2023

[5] Fujita S, Sumi M, Tatsukawa E, Nagano K, Katase N. Expressions of extracellular matrix-remodeling factors in lymph nodes from oral cancer patients. Oral Diseases. 2020;**26**(7):1424-1431. DOI: 10.1111/odi.13419. Epub 2020 Jun 4

[6] Higgins KM, Wang JR. State of head and neck surgical oncology research—A review and critical appraisal of landmark studies. Head & Neck. 2008;**30**(12):1636-1642. DOI: 10.1002/hed.20863

[7] Talmi YP, Takes RP, Alon EE, Nixon IJ, López F, de Bree R, et al. Prognostic value of lymph node ratio in head and neck squamous cell carcinoma. Head & Neck. 2018;**40**(5):1082-1090. DOI: 10.1002/hed.25080. Epub 2018 Feb 2

[8] Rinaldo A, Devaney KO, Ferlito A. Immunohistochemical studies in the identification of lymph node micrometastases in patients with squamous cell carcinoma of the head and neck. ORL: Journal for Otorhinolaryngology and Its Related Specialties. 2004;**66**(1):38-41. DOI: 10.1159/000077232

[9] West H, Jin J. Lymph nodes and lymphadenopathy in cancer. JAMA Oncology. 2016;**2**(7):971. DOI: 10.1001/jamaoncol.2015.3509

[10] Ferrer R. Lymphadenopathy: Differential diagnosis and evaluation. American Family Physician. 1998;**58**(6):1313-1320

[11] Karaman S, Detmar M. Mechanisms of lymphatic metastasis. The Journal of Clinical Investigation. 2014;**124**(3):922-928. DOI: 10.1172/JCI71606. Epub 2014 Mar 3

[12] Vuity D, McMahon J, Hislop S, McCaul J, Wales C, Ansell M, et al. Sentinel lymph node biopsy for early oral cancer—Accuracy and considerations in patient selection. The British Journal of Oral & Maxillofacial Surgery. 2022;**60**(6):830-836. DOI: 10.1016/j.bjoms.2021.12.058. Epub 2022 Jan 12

[13] de Bree R, de Keizer B, Civantos FJ, Takes RP, Rodrigo JP, Hernandez-Prera JC, et al. What is the role of sentinel lymph node biopsy in the management of oral cancer in 2020? European Archives of Oto-Rhino-Laryngology. 2021;**278**(9):3181-3191. DOI: 10.1007/s00405-020-06538-y. Epub 2020 Dec 28

[14] Detmar M, Hirakawa S. The formation of lymphatic vessels and its importance in the setting of malignancy. The Journal of Experimental Medicine. 2002;**196**(6):713-718. DOI: 10.1084/jem.20021346

[15] Meier F, Will S, Ellwanger U, Schlagenhauff B, Schittek B, Rassner G, et al. Metastatic pathways and time courses in the orderly progression of cutaneous melanoma. The British Journal of Dermatology. 2002;**147**(1):62-70. DOI: 10.1046/j.1365-2133.2002.04867. x

[16] Lengelé B, Hamoir M, Scalliet P, Grégoire V. Anatomical bases for the radiological delineation of lymph node areas. Major collecting trunks, head and neck. Radiotherapy and Oncology. 2007;**85**(1):146-155. DOI: 10.1016/j.radonc.2007.02.009. Epub 2007 Mar 23

[17] Bennett SH, Futrell JW, Roth JA, Hoye RC, Ketcham AS. Prognostic significance of histologic host response in cancer of the larynx or hypopharynx. Cancer. 1971;**28**(5):1255-1265. DOI: 10.1002/1097-0142(1971)28:5<1255:aid-cncr2820280524>3.0.co;2-a

[18] Shah JP. Patterns of cervical lymph node metastasis from squamous carcinomas of the upper aerodigestive tract. American Journal of Surgery. 1990;**160**(4):405-409. DOI: 10.1016/s0002-9610(05)80554-9

[19] Wang Y, Ow TJ, Myers JN. Pathways for cervical metastasis in malignant neoplasms of the head and neck region. Clinical Anatomy. 2012;**25**(1):54-71. DOI: 10.1002/ca.21249. Epub 2011 Aug 18

[20] Wang Y, Oliver G. Current views on the function of the lymphatic vasculature in health and disease. Genes & Development. 2010;**24**(19):2115-2126. DOI: 10.1101/gad.1955910

[21] Mescher AL. Junqueira's Basic Histology: Text and Atlas. 12th ed. New York: McGraw-Hill; 2010. pp. 241-244

[22] Carr I. Lymphatic metastasis. Cancer Metastasis Reviews. 1983;**2**(3):307-317. DOI: 10.1007/BF00048483

[23] Barnes L, Fan C. Pathology of the Head and Neck: Basic Considerations and New Concepts. 4th ed. Philadelphia: W.B. Saunders; 2003

[24] Wong SY, Hynes RO. Lymphatic or hematogenous dissemination: How does a metastatic tumor cell decide? Cell Cycle. 2006;**5**(8):812-817. DOI: 10.4161/cc.5.8.2646. Epub 2006 Apr 17

[25] Djalilian M, Weiland LH, Devine KD, Beahrs OH. Significance of jugular vein invasion by metastatic carcinoma in radical neck dissection. American Journal of Surgery. 1973;**126**(4):566-569. DOI: 10.1016/s0002-9610(73)80050-9

[26] Dias FL, Lima RA, Kligerman J, Farias TP, Soares JR, Manfro G, et al. Relevance of skip metastases for squamous cell carcinoma of the oral tongue and the floor of the mouth. Otolaryngology and Head and Neck Surgery. 2006;**134**(3):460-465. DOI: 10.1016/j.otohns.2005.09.025

[27] Bittar RF, Ferraro HP, Ribas MH, Lehn CN. Predictive factors of occult neck metastasis in patients with oral squamous cell carcinoma. Brazilian Journal of Otorhinolaryngology. 2016;**82**(5):543-547. DOI: 10.1016/j.bjorl.2015.09.005. Epub 2015 Dec 17

[28] Cho JH, Lee YS, Sun DI, Kim MS, Cho KJ, Nam IC, et al. Prognostic impact of lymph node micrometastasis in oral and oropharyngeal squamous cell carcinomas. Head & Neck. 2016;**38**(Suppl. 1):E1777-E1782. DOI: 10.1002/hed.24314. Epub 2015 Dec 17

[29] Ferlito A, Shaha AR, Rinaldo A. The incidence of lymph node micrometastases in patients pathologically staged N0 in cancer of oral cavity and oropharynx. Oral Oncology. 2002;**38**(1):3-5. DOI: 10.1016/s1368-8375(01)00037-9

[30] Zirk M, Safi AF, Buller J, Nickenig HJ, Dreiseidler T, Zinser M, et al. Lymph node ratio as prognosticator in floor of mouth squamous cell carcinoma patients. Journal of Cranio-Maxillo-Facial Surgery. 2018;**46**(2):195-200. DOI: 10.1016/j.jcms.2017.11.021. Epub 2017 Nov 26

[31] Huang TH, Li KY, Choi WS. Lymph node ratio as prognostic variable in oral squamous cell carcinomas: Systematic review and meta-analysis. Oral Oncology. 2019;**89**:133-143. DOI: 10.1016/j.oraloncology.2018.12.032. Epub 2019 Jan 8

[32] Iocca O, Di Maio P, De Virgilio A, Pellini R, Golusiński P, Petruzzi G, et al. Lymph node yield and lymph node ratio in oral cavity and oropharyngeal carcinoma: Preliminary results from a prospective, multicenter, international cohort. Oral Oncology. 2020;**107**:104740. DOI: 10.1016/j.oraloncology.2020.104740. Epub 2020 May 4

[33] Arun I, Maity N, Hameed S, Jain PV, Manikantan K, Sharan R, et al. Lymph node characteristics and their prognostic significance in oral squamous cell carcinoma. Head & Neck. 2021;**43**(2):520-533. DOI: 10.1002/hed.26499. Epub 2020 Oct 6

[34] Roberts TJ, Colevas AD, Hara W, Holsinger FC, Oakley-Girvan I, Divi V. Number of positive nodes is superior to the lymph node ratio and American joint committee on cancer N staging for the prognosis of surgically treated head and neck squamous cell carcinomas. Cancer. 2016;**122**(9):1388-1397. DOI: 10.1002/cncr.29932. Epub 2016 Mar 11

[35] Mamic M, Luksic I. Lymph node characteristics and their prognostic significance in oral squamous cell carcinoma. Head & Neck. 2021;**43**(8):2554-2555. DOI: 10.1002/hed.26715. Epub 2021 May 8

[36] Lydiatt WM, Patel SG, O'Sullivan B, Brandwein MS, Ridge JA, Migliacci JC, et al. Head and neck cancers-major changes in the American joint committee on cancer eighth edition cancer staging manual. CA: a Cancer Journal for Clinicians. 2017;**67**(2):122-137. DOI: 10.3322/caac.21389. Epub 2017 Jan 27

[37] Voss JO, Freund L, Neumann F, Mrosk F, Rubarth K, Kreutzer K, et al. Prognostic value of lymph node involvement in oral squamous cell carcinoma. Clinical Oral Investigations. 2022;**26**(11):6711-6720. DOI: 10.1007/s00784-022-04630-7. Epub 2022 Jul 27

Chapter 4

Lymph Node Yield and Ratio during Surgery for Advanced Laryngeal Carcinoma

Ahmed S. Elhamshary, Mostafa I. Ammar,
Eslam Farid Abu Shady and Ahmed Elnaggar

Abstract

Lymph node metastasis represents one of the most important prognostic factors in patients with head and neck squamous cell carcinomas (HNSCC). Lymph node yield (LNY) is the term used to indicate the total number of dissected lymph nodes following neck dissection, while lymph node ratio (LNR) is the proportion of metastatic lymph nodes to the total number of removed lymph nodes following neck dissection. This ratio serves to determine both the extent of cancer lymphatic spread and the effectiveness of its clearance. Calculating LNY and LNR following neck dissection holds particular significance when dealing with advanced laryngeal cancer. These values are supposed to have a direct impact on both prognosis and oncological outcomes, warranting their inclusion in the staging of such patients. Wide variations were observed in both LNY and LNR, which were mainly dependent not only on the tumor burden but also on surgical and pathological skills. Therefore, standardization is required in the pathological processing as well as surgical techniques of neck dissections to minimize these variations. Further studies are needed to validate these observations and to guide their inclusion in pathological TNM classification.

Keywords: lymph node ratio, glottic carcinoma, lymph node yield, neck dissection, laryngeal carcinoma

1. Introduction

1.1 The anatomy and morphology of cervical lymph nodes

The neck is characterized by the presence of a rich plexus of lymph nodes and channels. In the 1930s, the Memorial Sloan-Kettering Cancer Center designed a classification system of cervical lymph nodes groups into various anatomic levels. This system, initially used for labeling neck dissection specimens, since then gained a worldwide acceptance. A large study in the same center adopting this classification system is considered a landmark study describing the pattern of lymphatic metastasis of different head and neck primaries and the lymph node groups at high risk of metastasis for each primary [1].

IntechOpen

In 1991, the American Head and Neck Society subsequently adopted that system [2], and then the classification system was revised in 2002 [3]. This scheme is now widely implemented as a "common language" between clinicians involved in the care of head and neck cancer patients to describe and report lymphatic metastasis status [4]. In this classification, the neck is divided into six levels (**Figure 1** and **Table 1**).

Patterns of lymphatic flow and region-specific lymphatic drainage:

There are approximately 150 lymph nodes on either side of the neck. The normal range in size is from 3 mm to 3 cm, but most nodes are less than a centimeter. Within level II, the largest node is often called the jugulodigastric node and is situated within the triangle formed by the internal jugular vein, facial vein, and posterior belly of the digastric muscle. It is important because it receives lymph from a wide area, which includes the submandibular region, the oropharynx, palatine tonsils, and oral cavity. The jugulo-omohyoid nodes are situated at the junction between the middle and lower cervical group (low level III/high level IV) where the omohyoid muscle crosses the internal jugular vein and receives lymph from a wide area, which includes the anterior floor of mouth, oropharynx, and larynx.

In the neck lymphatic flow follows an orderly and predictable path. Through studying metastasis pattern of squamous-cell carcinoma of the larynx and hypopharynx prior to surgical therapy. In 1964, Fisch [5] studied the patterns of cervical lymphatic flow by injecting oil-based contrast media into the post-auricular lymphatics followed by lymphography, the flow of contrast was from post-auricular lymphatics to nodes just below and behind the angle of the mandible (highest Level IIB or VA nodes) called the junctional nodes then contrast flowed to nodes along the spinal accessory nerve posteriorly and along the jugular nodes anteriorly. Contrast in the posterior triangle then flowed to transverse cervical nodes (level VB) and then

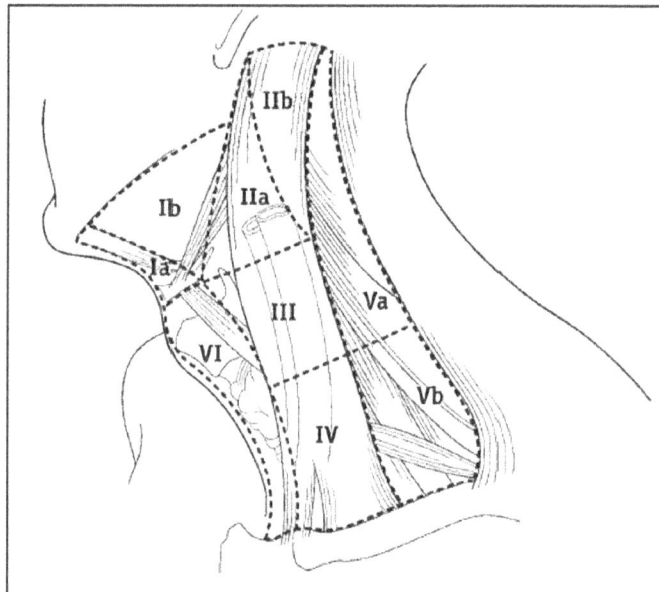

Figure 1.
Lymph node levels and sublevels of the neck (adapted from Givi and Andersen [4]).

Border				
Level	Superior	Inferior	Anterior (medial)	Posterior (lateral)
IA	Symphysis of mandible	Body of hyoid	Anterior belly of contralateral digastric muscle	Anterior belly of ipsilateral digastric muscle
IB	Body of mandible	Posterior belly of digastric muscle	Anterior belly of digastric muscle	Stylohyoid muscle
IIA	Skull base	Horizontal plane defined by the inferior body of the hyoid bone	Stylohyoid muscle	Vertical plane defined by the spinal accessory nerve
IIB	Skull base	Horizontal plane defined by the inferior body of the hyoid bone	Vertical plane defined by the spinal accessory nerve	Posterior border of the sternocleidomastoid muscle
III	Horizontal plane defined by the inferior body of the hyoid	Horizontal plane defined by the inferior border of the cricoid cartilage	Lateral border of the sternohyoid muscle	Posterior border of the sternocleidomastoid muscle
IV	Horizontal plane defined by the inferior border of the cricoid cartilage	Clavicle	Lateral border of the sternohyoid muscle	Posterior border of the sternocleidomastoid muscle
VA	Apex of the convergence of the sternocleidomastoid and trapezius muscle	Horizontal plane defined by the inferior border of the cricoid cartilage	Posterior border of the sternocleidomastoid muscle or sensory branches of cervical plexus	Anterior border of the trapezius muscle
VB	Horizontal plane defined by the inferior border of the cricoid cartilage	Clavicle	Posterior border of the Sternocleidomastoid muscle or sensory branches of cervical plexus	Anterior border of the trapezius muscle
VI	Hyoid bone	Suprasternal notch	Common carotid artery	Common carotid artery
VII	Suprasternal notch	Innominate artery	Common carotid artery	Common carotid artery

Table 1.
Anatomic boundaries of the lymph node levels of the neck [4].

medially to low jugular nodes (level IV). The contrast from spinal accessory nodes flowed to jugular nodes anteriorly but never moved in the opposite direction. There was also no contralateral or retrograde lymphatic flow. This important observation is the basis of the notion that metastases to level V nodes are extremely rare and this means that junctional nodes are not involved in the majority of cancer (**Figure 2**) [5].

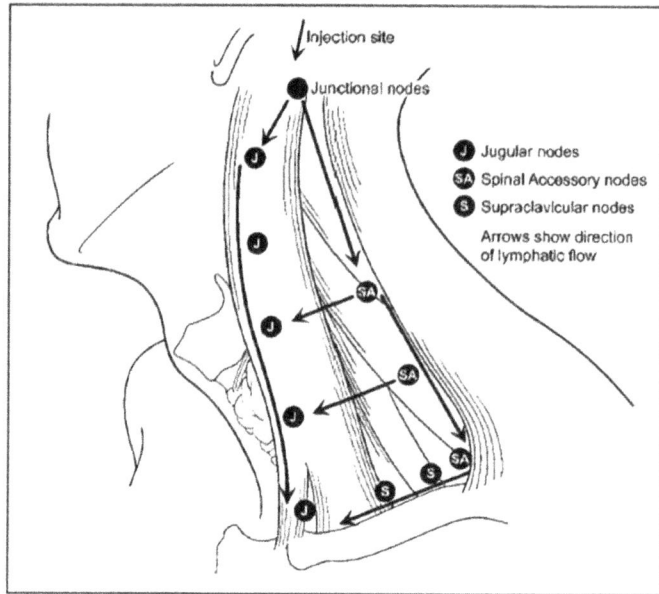

Figure 2.
Patterns of lymphatic flow based on Fisch experiments (adapted from Givi and Andersen [4]).

It is important to realize that contralateral neck spread may occur early in those tumors situated in or near the midline.

A landmark study by Lindberg [6] reviewed 2044 patients with HNSCC of oral cavity, oropharynx, nasopharynx, supraglottis, and hypopharynx and published the topographical distribution of clinically evident cervical metastases was set out. Byers et al. [7], from MD Anderson Cancer Center, identified these distinct patterns of spread to the neck based on the primary site relying on patterns of nodal metastasis among 428 patients with HNSCC.

The laryngeal lymphatic drainage is separated into upper and lower systems based on its embryological origins, with a division that occurs at the level of the true vocal cord. The supraglottis drains through vessels that accompany the superior laryngeal pedicle *via* the thyroid membrane to reach levels II/III, with a greater tendency for bilateral nodal drainage. The lower system drains directly into levels III/IV through vessels that pass through or behind the cricothyroid membrane and also into the prelaryngeal, pretracheal, or paratracheal nodes (level VI), before reaching the deep cervical nodes. Because the vocal cords are relatively avascular, they have a sparse lymphatic drainage, hence lymph node metastases from small carcinomas at this site are uncommon [6, 7].

1.2 Cervical lymph node metastasis in advanced laryngeal carcinoma

Laryngeal cancer constitutes about one-third of all head and neck cancers [8]. Its incidence continues to increase in South-East Asia, Africa, and the Western Pacific, while the incidence is declining in Western countries. Heavy and prolonged smoking and alcohol ingestion are the main risk factors for laryngeal cancer. Other risk factors include laryngopharyngeal reflux disease, occupational exposure to solvents, sulfuric acid, asbestos, and human papillomavirus (HPV) infection [9].

Cervical lymph node metastases are one of the most adverse prognostic factors in squamous cell carcinoma (SCC) of the larynx and are always associated with poor oncological outcomes. Despite advances in diagnostic imaging, currently, there are no imaging modalities capable of accurately detecting occult disease in the clinically negative neck (cN0). Neck dissection remains the gold standard for nodal staging in cN0 patients with high-risk HNSCC if the risk of neck metastasis is more than 15–20%. The accuracy of surgical staging in the cN0 neck depends on the extent of neck dissection as well as on the scrutiny of histopathologic examination for detecting occult metastatic disease. Therefore, the probability of identifying metastasis in lymph nodes relies on the skills of both surgeons and pathologists [10].

In the eighth edition of the pathological pTNM classification, the pN classification of neck dissection in laryngeal carcinomas is based on the size and laterality, presence of single or multiple metastatic lymph nodes, and presence of extracapsular spread (ECS) in metastatic lymph nodes [11].

However, several studies have highlighted some serious limitations in pN classification. First, it does not take into consideration the number of metastatic lymph nodes in the neck. In a study on 8351 hypopharyngeal and laryngeal cancer patients, the overall survival is inversely proportionate to the number of metastatic lymph nodes in neck dissection specimens, while the node size or contralateral lymph node involvement did not have a statistically significant impact on survival. Therefore, the authors recommended the incorporation of quantitative metastatic nodal burden in nodal classification for laryngeal and hypopharyngeal cancers to improve its prognostic value and better identify the need for adjuvant treatment [11]. Second, this pN classification does not assess the number of removed lymph nodes in the neck dissection specimen. Many studies have proved that a higher number of dissected lymph nodes in the neck dissection specimen was associated with improved survival in HNSCC patients [12, 13].

The accumulating evidence in support of the prognostic value of both the number of dissected lymph nodes and the number of metastatic lymph nodes in the neck dissection specimen has inspired many head and neck surgeons to further investigate their prognostic capacity, standardize their terminology, and discover their uses and limitations.

2. Aim of this chapter

This is a review chapter to illustrate the prognostic importance of lymph node yield and ratio during surgery for advanced laryngeal carcinoma as well as their uses and limitations.

3. Lymph node yield (LNY)

3.1 Definition

LNY refers to the total number of removed lymph nodes following neck dissections [14]. LNY was first described by Agrama et al. in modified radical neck dissection (MRND) for HNSCC, showing variations in LNY values [15].

3.2 Prognostic value

There has been a growing interest in LNY among head and neck surgeons in the last decades. Ebrahimi et al. [16] were the first to try to study the prognostic value of LNY in 225 patients undergoing elective neck dissection for oral SCC. This single institutional study concluded that nodal yield <18 was significantly associated with decreased overall survival, disease-specific survival, and disease-free survival. In a larger study cohort of 1567 cN0 oral SCC patients from nine international cancer centers, Ebrahimi et al. [17] confirmed the same finding, proving that LNY is a strong independent prognosticator for decreased survival and increased risk of locoregional recurrence. In a study of 4341 patients with pN0 oral SCC undergoing elective neck dissection, Lemieux et al. [18] found that greater LNY of more than 22 nodes was significantly associated with increased overall survival. Two studies in HNSCC patients, including node-positive patients of all primary sites, have shown that LNY ≥ 18 was associated with improved overall survival and reduced risk of locoregional failure [19, 20].

In laryngeal SCC patients, some studies found that LNY was not significantly correlated with the overall survival and disease-free survival. These studies had a retrospective study design with a small sample size [21–23]. In one of them, the neck dissection was done by different surgeons, and lymph node examination was performed by different histopathologists, making more bias [22]. Additional prospective studies with larger sample sizes are required to accurately define the minimum LNY and verify its prognostic capacity in patients with laryngeal SCC.

LNY is an objective tool for surgical adequacy of neck dissection as well as pathological processing of neck dissection specimens, particularly when standardization of the surgical techniques and pathological processing will be necessary to allow reproducibility and statistical comparison of similar patient groups [24]. Moreover, a review has shown the volume of neck dissection specimen is an indicator of surgical expertise that could be used to assess trainees' progress and for quality maintenance in large head and neck centers [25].

3.3 Variations in LNY

The surgeon should dissect a significant number of lymph nodes from the neck levels as possible, to approach the average count of dissected lymph nodes as close to the average lymph node count as possible [25]. In a cadaveric study, there were an average of 20 lymph nodes in supraomohyoid neck dissection compared to an average of 30 in lateral neck dissection [26]. Similar LNY was retrieved in a retrospective review of 414 patients undergoing therapeutic neck dissection, obtaining mean LNY of 21.7 in supraomohyoid neck dissection compared to 27.1 in lateral neck dissection [27]. Preoperative CT-planned neck level volume estimates correlate with neck dissection specimen volume but did not correlate with LNY [28].

According to the 8th edition of the TNM classification, pN pathological classification requires a selective neck dissection specimen to include 10 or more lymph nodes, while a radical or modified radical neck dissection specimen should encompass 15 or more lymph nodes [11]. However, the minimum LNY in patients with laryngeal SCC was quite variable (35.9–12) with an average of 24 for selective level II–IV neck dissections [22, 24].

Norling et al. [29] have compared LNY values in cadaveric and clinical selective neck dissections (SND). Based on literature review, the minimum average LNY was

19.4 in clinical supraomohyoid SND (levels I–III) and 26.4 in clinical lateral SND (levels II–IV), while the lowest mean LNY was 8.8 in cadaveric supraomohoyid SND (levels I–III) and 10.4 LNs in cadaveric lateral SND (levels II–IV) [29]. Four clinical studies reported the mean nodal yield for levels I, II, III, IV, and V. Means for level I were 5.7, 5.27, and 4.0, for level II: 12.6, 11.2, 11.2, and 9.43, for level III: 8.49, 7.6, 7.6 and 7.2, for level IV: 8.7, 7.43, 7.3 and 6.9, and for level V: 9.7 and 9.02 [29, 30]. In a study of LNY in SND (levels II–IV) in 45 advanced laryngeal cancer patients, the mean LNY for level II was 9.1, for level III: 11.5, and for level IV: 6.2 with level III containing most lymph nodes. The mean LNY of SND (levels II–IV) was 47.7 [31].

Over the last decades, there have been wide variations in reported LNY values in many studies because some studies were conducted on patients with different tumor locations together with the lack of uniformity in the proper extent of neck dissection between different centers and surgeons, making it difficult to set the standards for LNY values in different neck dissection types. Moreover, the current literature has moved towards superselective neck dissection, adding more difficulty. For this reason, separating neck dissection specimens into the individual neck levels before being sent for the histopathological examination should be recommended in the future guidelines to establish standard LNY values in different neck levels and, therefore, to assess the adequacy of neck dissection [30, 32].

3.4 Factors influencing LNY

Many factors could affect the nodal yield in neck dissections for different head and neck cancers. (1) Patient and tumor factors: a univariate analysis of patient and tumor factors in oral SCC revealed that women, older people, BMI <25, small (T1, T2) tumors, absent positive lymph nodes, and absent perineural invasion were significantly associated with reduced LNY. However, the multivariate analysis of significant factors found that older age and BMI <25 are the most significant patient factors, while T classification and the presence of positive lymph nodes are the most significant tumor factors [33, 34]. In mucosal HNSCC, p16-positive tumors have yielded around 2.4 more lymph nodes than their p16-negative tumors. HPV status significantly affected LNY, particularly in oral SCC [30]. (2) Pathological factors: A newly developed pathology protocol with examination of residual fibrofatty tissue in neck dissection specimen has significantly increased LNY [34]. (3) Treatment factors include surgical factors or preoperative or postoperative radiation. Both standard surgical technique and surgical experience had a significant impact on nodal yield in neck dissections. The horizontal neck dissection using standard fascia unwrapping technique was associated with a significantly superior nodal yield in levels I, II, III, and IV, and in overall nodal yield than the vertical neck dissection in the control group [24]. Level-by-level neck dissection resulted in a statistically significant higher LNY than en bloc or monoblock neck dissection in both selective neck dissection and individual neck levels [32]. Surgical experience influences not only LNY in neck dissections but also oncological outcomes with more recurrences occurring with less-experienced surgeons [25]. Both preoperative radiotherapy and chemoradiotherapy over the head and neck region reduce the nodal yield significantly in neck dissections, compared with patients who did not receive radiotherapy or who received postoperative radiotherapy [21, 27, 34–37].

4. Lymph node ratio (LNR)

4.1 Definition

Lymph node ratio (LNR) or lymph node density (LND) is defined as the ratio of the metastatic lymph nodes to the total number of dissected lymph nodes following neck dissection. This ratio determines the extent of cancer lymphatic spread and extent of its clearance. Both LNR and LND are being used interchangeably in the current literature [14]. For laryngeal and hypopharyngeal SCC, LNR ranges from 0.03 to 0.14 [38].

The standard histopathological examination of lymph nodes in neck dissection specimens starts with making a single, longitudinal section through the center of each lymph node followed by Hematoxylin and Eosin (H&E) staining and examination of the section under light microscopy for metastatic deposits. This standard technique is widely used [39]. Serial section H & E staining at 3–4 mm of the lymph node and cytokeratin immunohistochemical analysis demonstrated a higher detection rate of micrometastatic disease in pN0 specimens [40].

4.2 Prognostic value

In their extensive multi-institutional study involving 4254 patients with oral SCC, Patel et al. [30, 41] reported that higher LND was significantly associated with lower overall survival, disease-specific survival, disease-free survival, and higher rates of locoregional and distant metastases. Comparing a newly developed TNM staging based on LND with the standard TNM staging revealed that the new TNM staging was superior to the standard TNM staging in all survival parameters [41].

In laryngeal and hypopharyngeal SCC patients, a meta-analysis found that higher LNR values were significantly associated with shorter overall survival, disease-specific survival, and disease-free survival [38]. Several studies have reported similar results and proved a significant association between increased LNR and higher risk for locoregional and distant recurrences in patients with advanced laryngeal or hypopharyngeal SCC, which should be considered for adjuvant treatment. Therefore, LNR could be used for risk stratification of laryngeal SCC patients as well as adjuvant treatment planning and follow-up [42–47].

Wang et al. [47] investigated the potential role of LNR in the staging of laryngeal SCC. They gathered data from the SEER database, encompassing 1963 patients, and supplemented it with an additional 27 patients from their own institution for validation. By determining optimal LNR cutoff values, the patients were categorized into three risk groups based on LNR values: ≤0.09, 0.09–0.20, and >0.20, which were significantly different in disease-specific survival and overall survival. Therefore, the authors recommended that incorporating LNR in N classification could enhance the staging process.

However, two studies found that LNR was not significantly correlated with the overall survival and disease-free survival in laryngeal SCC patients. These studies had a retrospective study design with a small sample size [21, 22]. In one of them, the neck dissection was done by different surgeons, and lymph node examination was performed by different histopathologists, making more bias [22].

4.3 Factors affecting LNR in neck dissections

Three main factors can potentially affect LNR and nodal staging as follows:
(1) tumor factors (the extent of neck involvement): the number of metastatic lymph

nodes in neck specimens directly influences LNR; (2) surgical factors (the extent of neck dissection): the total number of dissected lymph nodes in neck dissection specimens varies significantly in different neck dissection types, with modified radical neck dissection producing the largest number of LNs (around 34), followed by SND level II–V (around 23), SND level I–III (about 18), and SND II–IV (about 17). With the current trend towards superselective neck dissection, LNR is expected to vary significantly in more limited neck dissections [13]. Therefore, further studies are required to define the cutoff points for LNR in different types of neck dissection; (3) pathological factors (the accuracy of the histopathological examination): variations in harvesting higher number of lymph nodes from neck specimens by different pathologists as well as increased detection of lymph node micrometastases by serial sectioning, immunohistochemistry, and/or the use of molecular techniques were proved to be associated with changes in the value of LNR [13, 36, 48]. Therefore, LNR is a reflection not only of disease burden but also of surgical and pathological quality standards. This clearly highlights the need for standardized protocols in the processing of the neck dissection specimens as well as surgical practice [49].

4.4 Limitations

Some limitations might affect the value of the LNR. First, LNR does not include the prognostic information associated with the presence of metastatic lymph nodes with ECS. In a retrospective study on 1190 patients with HNSCC, the number of metastatic lymph nodes with ECS was significantly related to the disease-specific survival [50]. Second, for pN0 patients, the value of LNR regardless of the number of dissected lymph nodes is always 0%, leading to loss of information on the nodal yield in this group of patients. Many studies have proved that an increased lymph node yield in pN0 patients resulted in significant survival improvement [18, 51].

4.5 Weighted lymph node ratio (WLNR)

The weighted lymph node ratio (WLNR) is a newly developed equation designed to integrate predictive data related to the number of metastatic lymph nodes with ECS and LNY for pN0 patients to LNR. The calculation of WLNR is outlined in the following equation: WLNR = [((number of positive lymph nodes without ECS × 1.054) + (number of positive lymph nodes with ECS × 1.199) + 0.5)/(total number of dissected lymph nodes +0.5)] × 100 [52].

Using the WLNR value, HNSCC patients are categorized into four groups, which exhibit notable differences in 5-year disease-specific survival rates as we compare one category to the next. The groups are as follows: Category I (WLNR ≤ 3.4%); Category II (WLNR ranging from 3.5% to 7.4%); Category III (WLNR ranging from 7.5% to 15.4%); and Category IV (WLNR ≥ 15.5%) [52].

Compared with the TNM pN classification (pN0, pN1, pN2, pN3), the WLNR classification was more reliable in predicting disease-specific survival along with both regional and distant recurrence-free survival. This classification improves the prognostic accuracy of the eighth edition of the pTNM classification and serves as a valuable resource for assessing the postoperative staging of the neck dissections in HPV-negative HNSCC patients [52]. In a study of 197 HNSCC patients with regional recurrence treated with salvage neck dissection, WLNR had proved to provide a significant prognostic capacity for disease-specific survival [53].

5. Conclusions

The estimation of LNY and LNR following neck dissection holds significant importance, particularly in advanced laryngeal cancer because they have a direct impact on both the prognosis and the oncological outcomes of such patients. Therefore, incorporation of LNR in the nodal staging of advanced laryngeal cancer is needed, once its prognostic capacity is well studied. Wide variations in both LNY and LNR were mainly affected by tumor, surgical, and pathological factors. Therefore, standardized protocols are needed in the pathological processing and surgical techniques of neck dissections to minimize these variations. Some limitations were observed with the use of LNR, leading to the introduction of weighted LNR. More well-designed research with larger samples is required to verify the reliability of LNY and LNR and clearly define their uses and limitations in the future.

Acknowledgements

No dedicated funding was provided for this study by any external source, whether governmental, commercial, or non-profit.

Conflict of interest

The authors declare no conflict of interest.

Notes/thanks/other declarations

This study did not receive any dedicated funding from external organizations, whether they are commercial enterprises or non-profit sectors. The authors have stated that there are no conflicts of interest.

Acronyms and abbreviations

HNSCC	head and neck squamous cell carcinoma
LNY	lymph node yield
LNR	lymph node ratio
HPV	human papilloma virus
SCC	squamous cell carcinoma
cN0	clinically negative neck
MRND	modified radical neck dissection
ECS	extracapsular spread
pN0	pathologically negative neck
SND	selective neck dissection
RND	radical neck dissection
UICC	International Cancer Control
AJCC	American Joint Committee on Cancer
WLNR	weighted lymph node ratio

Author details

Ahmed S. Elhamshary[1], Mostafa I. Ammar[1], Eslam Farid Abu Shady[2]*
and Ahmed Elnaggar[1]

1 Otorhinolaryngology-Head and Neck Surgry Department, Tanta University, Tanta,
Egypt

2 Otolaryngology-Head and Neck Surgry Department, Benha University, Benha,
Egypt

*Address all correspondence to: eslam.farid@fmed.bu.edu.eg

IntechOpen

References

[1] Patel KN, Shah JP. Neck dissection: Past, present, future. Surgical Oncology Clinics. 2005;**14**(3):461-477

[2] Robbins KT, Medina JE, Wolfe GT, Levine PA, Sessions RB, Pruet CW. Standardizing neck dissection terminology: Official report of the Academy's Committee for Head and Neck Surgery and Oncology. Archives of Otolaryngology–Head & Neck Surgery. 1991;**117**(6):601-605

[3] Robbins KT, Clayman G, Levine PA, Medina J, Sessions R, Shaha A, et al. Neck dissection classification update: Revisions proposed by the American Head and Neck Society and the American Academy of Otolaryngology–Head and Neck Surgery. Archives of Otolaryngology–Head & Neck Surgery. 2002;**128**(7):751-758

[4] Givi B, Andersen PE. Rationale for modifying neck dissection. Journal of Surgical Oncology. 2008;**97**(8):674-682

[5] Fisch U. Cervical lymphography in cases of laryngo-pharyngeal carcinoma. The Journal of Laryngology & Otology. 1964;**78**(8):715-726

[6] Lindberg R. Distribution of cervical lymph node metastases from squamous cell carcinoma of the upper respiratory and digestive tracts. Cancer. 1972;**29**(6):1446-1449

[7] Byers RM, Wolf PF, Ballantyne AJ. Rationale for elective modified neck dissection. Head & Neck Surgery. 1988;**10**(3):160-167

[8] Koroulakis A, Agarwal M. Laryngeal cancer. Available from: https://www.statpearls.com/point-of-care/24035 [Accessed: August 15, 2023]

[9] Nocini R, Molteni G, Mattiuzzi C, Lippi G. Updates on larynx cancer epidemiology. Chinese Journal of Cancer Research. 2020;**32**(1):18-25

[10] Pou JD, Barton BM, Lawlor CM, Frederick CH, Moore BA, Hasney CP. Minimum lymph node yield in elective level I–III neck dissection. The Laryngoscope. 2017;**127**(9):2070-2073

[11] Brierley JD, Gospodarowicz MK, Ch W, editors. International Union against Cancer (UICC). TNM Classification of Malignant Tumors. 8th ed. Wiley-Blackwell: Oxford, UK; 2017

[12] Ho AS, Kim S, Tighiouart M, Gudino C, Mita A, Scher KS, et al. Association of quantitative metastatic lymph node burden with survival in hypopharyngeal and laryngeal cancer. JAMA Oncology. 2018;**4**(7):985-989

[13] de Kort WWB, Maas SLN, Van Es RJJ, Willems SM. Prognostic value of the nodal yield in head and neck squamous cell carcinoma: A systematic review. Head & Neck. 2019;**41**:2801-2810

[14] Iocca O, Di Maio P, De Virgilio A, Pellini R, Golusiński P, Petruzzi G, et al. Lymph node yield and lymph node ratio in oral cavity and oropharyngeal carcinoma: Preliminary results from a prospective, multicenter, international cohort. Oral Oncology. 2020;**107**:104740

[15] Agrama MT, Reiter D, Topham AK, Keane WM. Node counts in neck dissection: Are they useful in outcomes research? Otolaryngology and Head and Neck Surgery. 2001;**124**(4):433-435

[16] Ebrahimi A, Zhang WJ, Gao K, Clark JR. Nodal yield and survival in oral

squamous cancer: Defining the standard of care. Cancer. 2011;**117**(13):2917-2925

[17] Ebrahimi A, Clark JR, Amit M, Yen TC, Liao CT, Kowalski LP, et al. Minimum nodal yield in oral squamous cell carcinoma: Defining the standard of care in a multicenter international pooled validation study. Annals of Surgical Oncology. 2014;**21**(9):3049-3055

[18] Lemieux A, Kedarisetty S, Raju S, Orosco R, Coffey C. Lymph node yield as a predictor of survival in pathologically node negative oral cavity carcinoma. Otolaryngology and Head and Neck Surgery. 2016;**154**(3):465-472

[19] Divi V, Chen MM, Nussenbaum B, et al. Lymph node count from neck dissection predicts mortality in head and neck cancer. Journal of Clinical Oncology. 2016;**34**:3892-3897

[20] Divi V, Harris J, Harari PM, et al. Establishing quality indicators for neck dissection: Correlating the number of lymph nodes with oncologic outcomes (NRG Oncology RTOG 9501 and RTOG 0234). Cancer. 2016;**122**:3464-3471

[21] Topf MC, Philips R, Curry J, Magana LC, Tuluc M, Bar-Ad V, et al. Impact of lymph node yield in patients undergoing total laryngectomy and neck dissection. The Annals of Otology, Rhinology, and Laryngology. 2021;**130**(6):591-601

[22] Cayonu M, Tuna EU, Acar A, Dinc ASK, Sahin MM, Boynuegri S, et al. Lymph node yield and lymph node density for elective level II-IV neck dissections in laryngeal squamous cell carcinoma patients. European Archives of Oto-Rhino-Laryngology. 2019;**276**(10):2923-2927

[23] Bottcher A, Dommerich S, Sander S, Olze H, Stromberger C, Coordes A,

et al. Nodal yield of neck dissections and influence on outcome in laryngectomized patients. European Archives of Oto-Rhino-Laryngology. 2016;**273**(10):3321-3329

[24] Lörincz BB, Langwieder F, Möckelmann N, Sehner S, Knecht R. The impact of surgical technique on neck dissection nodal yield: Making a difference. European Archives of Oto-Rhino-Laryngology. 2016;**273**(5):1261-1267

[25] Morton RP, Gray L, Tandon DA, Izzard M, McIvor NP. Efficacy of neck dissection: Are surgical volumes important? Laryngoscope. 2009;**119**(6):1147-1152

[26] Friedman M, Lim JW, Dickey W, Tanyeri H, Kirshenbaum GL, Phadke DM, et al. Quantification of lymph nodes in selective neck dissection. The Laryngoscope. 1999;**109**(3):368-370

[27] Lippert D, Dang P, Cannon D, Harari PM, McCulloch TM, Hoffman MR. Lymph node yield in therapeutic neck dissection: Impact of dissection levels and prior radiotherapy. The Annals of Otology, Rhinology, and Laryngology. 2017;**126**(11):762-767

[28] Rees CA, Litchman JH, Wu X, Servos MM, Kerr DA, Halter RJ, et al. CT for estimating adequacy of lymph node dissection in patients with squamous cell carcinoma of the head and neck. Cancer Imaging. 2021;**21**(1):61

[29] Norling R, Therkildsen MH, Bradley PJ, Nielsen MB, Cv B. Nodal yield in selective neck dissection. Acta Oto-Laryngologica. 2013;**133**(9):965-971

[30] Lim RS, Evans L, George AP, de Alwis N, Stimpson P, Merriel S, et al. Do demographics and tumour-related factors affect nodal yield at neck

dissection? A retrospective cohort study. The Journal of Laryngology and Otology. 2017;**131**(S1):S36-S40

[31] Abdel Samea HAR, Elhamshary AAS, Abdelaziz MF, Abu Shady EF. Elective neck dissection during surgery for advanced glottic carcinoma with a clinically negative neck: Analysis of lymph node yield and early post-surgical outcomes. Benha Medical Journal. 2022;**39**(1):262-278

[32] Devaraja K, Pujary K, Ramaswamy B, Nayak DR, Kumar N, Nayak D. Lymph node yield in treatment naïve cases of head and neck squamous cell carcinoma: En bloc lymphadenectomy versus level-by-level dissection. The Journal of Laryngology and Otology. 2021;**135**:359-366

[33] Safi AF, Kauke M, Grandoch A, Nickenig HJ, Drebber U, Zöller J, et al. Clinicopathological parameters affecting nodal yields in patients with oral squamous cell carcinoma receiving selective neck dissection. Journal of Cranio-Maxillo-Facial Surgery. 2017;**45**(12):2092-2096

[34] Holcomb AJ, Perryman M, Goodwin S, Penn J, Villwock MR, Bur AM, et al. Pathology protocol increases lymph node yield in neck dissection for oral cavity squamous cell carcinoma. Head & Neck. 2020;**42**(10):2872-2879

[35] Johnstone PAS, Miller ED, Moore MG. Preoperative radiotherapy decreases lymph node yield of neck dissections for head and neck cancer. Otolaryngology and Head and Neck Surgery. 2012;**147**(2):278-280

[36] Marres C, De Ridder M, Hegger I, van Velthuysen M, Hauptmann M, Navran A, et al. The influence of nodal yield in neck dissections on lymph node ratio in

head and neck cancer. Oral Oncology. 2014;**50**(1):59-64

[37] Hintze J, Lang B, Subramaniam T, Kruseman N, O'Regan E, Brennan S, et al. Factors influencing nodal yield in neck dissections for head and neck malignancies. The Journal of Laryngology and Otology. 2023;**137**(8):925-929

[38] Abdeyrim A, He S, Zhang Y, Mamtali G, Asla A, Yusup M, et al. Prognostic value of lymph node ratio in laryngeal and hypopharyngeal squamous cell carcinoma: A systematic review and meta-analysis. Journal of Otolaryngology - Head & Neck Surgery. 2020;**49**(1):31

[39] Devaney KO, Rinaldo A, Ferlito A. Micrometastases in cervical lymph nodes from patients with squamous carcinoma of the head and neck: Should they be actively sought? Maybe. American Journal of Otolaryngology. 2007;**28**(4):271-274

[40] Barrera JE, Miller ME, Said S, Jafek BW, Campana JP, Shroyer KR. Detection of occult cervical micrometastases in patients with head and neck squamous cell cancer. The Laryngoscope. 2003;**113**(5):892-896

[41] Patel S, Amit M, Yen T, Liao C, Chaturvedi P, Agarwal J, et al. Lymph node density in oral cavity cancer: Results of the international consortium for outcomes research. British Journal of Cancer. 2013;**109**(8):2087-2095

[42] Grasl S, Janik S, Parzefall T, Formanek M, Grasl MC, Heiduschka G, et al. Lymph node ratio as a prognostic marker in advanced laryngeal and hypopharyngeal carcinoma after primary total laryngopharyngectomy. Clinical Otolaryngology. 2020;**45**:73-82

[43] Imre A, Pinar E, Dincer E, Ozkul Y, Aslan H, Songu M, et al. Lymph node

density in node-positive laryngeal carcinoma: Analysis of prognostic value for survival. Otolaryngology and Head and Neck Surgery. 2016;**155**(5):797-804

[44] Ryu IS, Roh JL, Cho KJ, Choi SH, Nam SY, Kim SY. Lymph node density as an independent predictor of cancer-specific mortality in patients with lymph node-positive laryngeal squamous cell carcinoma after laryngectomy. Head & Neck. 2015;**37**(9):1319-1325

[45] Petrarolha S, Dedivitis R, Matos L, Ramos D, Kulcsar M. Lymph node density as a predictive factor for worse outcomes in laryngeal cancer. European Archives of Oto-Rhino-Laryngology. 2020;**277**(3):833-840

[46] Kunzel J, Mantsopoulos K, Psychogios G, et al. Lymph node ratio is of limited value for the decision-making process in the treatment of patients with laryngeal cancer. European Archives of Oto-Rhino-Laryngology. 2015;**272**(2):453-461

[47] Wang YL, Li DS, Wang Y, Wang ZY, Ji QH. Lymph node ratio for postoperative staging of laryngeal squamous cell carcinoma with lymph node metastasis. PLoS One. 2014;**9**:e87037

[48] Negm H, Mosleh M, Fathy H, Hareedy A, Elbattawy A. Cytokeratin immunohistochemically detected nodal micrometastases in N0 laryngeal cancer: Impact on the overall occult metastases. European Archives of Oto-Rhino-Laryngology. 2013;**270**:1085-1092

[49] Talmi YP, Takes RP, Alon EE, Nixon IJ, López F, de Bree R, et al. Prognostic value of lymph node ratio in head and neck squamous cell carcinoma. Head & Neck. 2018;**40**:1082-1090

[50] de Juan J, García J, López M, Orús C, Esteller E, Quer M, et al. Inclusion of

extracapsular spread in the pTNM classification system: A proposal for patients with head and neck carcinoma. JAMA Otolaryngology. Head & Neck Surgery. 2013;**139**:483-488

[51] Merz S, Timmesfeld N, Stuck BA, Wiegand S. Impact of lymph node yield on outcome of patients with head and neck cancer and pN0 neck. Anticancer Research. 2018;**38**:5347-5350

[52] León X, Neumann E, Gutiérrez A, García J, López M, Quer M. Weighted lymph node ratio: New tool in the assessment of postoperative staging of the neck dissection in HPV-negative head and neck squamous cell carcinoma patients. Head & Neck. 2020;**42**:2912-2919

[53] Neumann ED, Sansa A, Casasayas M, Gutierrez A, Quer M, León X. Prognostic capacity of the weighted lymph node ratio in head and neck squamous cell carcinoma patients treated with salvage neck dissection. European Archives of Oto-Rhino-Laryngology. 2021;**278**(10):4005-4010

Chapter 5

Optimizing Nodal Staging in Intermediate and High-Risk Prostate Cancer: An Examination of Sentinel Lymph Node Dissection Using ICG/NIR

Robert M. Molchanov, Oleg B. Blyuss and Ruslan V. Duka

Abstract

This study evaluated the use of sentinel lymph node (SLN) dissection with indocyanine green/near-infrared (ICG/NIR) technology in laparoscopic radical prostatectomy for clinically localized prostate cancer (PCa). Conducted from 2020 to 2023, the study included 60 patients: 45 at intermediate or high risk underwent both SLN dissection and extended pelvic lymph node dissection (ePLND), while 15 low-risk patients had SLN dissection only. Sentinel nodes were identified in over 90% of cases. Body mass index (BMI) was found to influence the time taken to locate SLNs. Among intermediate and high-risk patients, 22% showed metastatic involvement. The procedure demonstrated a specificity of 90%, sensitivity of 80%, and positive predictive value of 95,7%. The study concludes that SLN dissection is a feasible and effective method for preoperative nodal staging in PCa, although further research is needed for optimization.

Keywords: sentinel lymph nodes, prostate cancer, laparoscopic radical prostatectomy, pelvic lymph node dissection, indocyanine green/near-infrared technology

1. Introduction

Introduced in the late twentieth century, sentinel lymph node biopsy (SLNB) has become a pivotal diagnostic and therapeutic instrument in oncology. It aims to pinpoint the first lymph nodes potentially affected by tumor metastasis, assisting in disease staging and shaping treatment choices. The principle behind SLNB is that if the sentinel node, being the first to encounter migrating cancer cells, test negative, a more invasive standard lymph node dissection may not be required. Currently, there is increasing evidence supporting the utility of SLNB as a predictive and therapeutic tool, with varying outcomes depending on the tumor location [1, 2].

One of the earliest studies focusing on sentinel lymph nodes was conducted in 1977 when Cabanas R.M. hypothesized that the sentinel lymph node could be the

IntechOpen

initial node affected during the progression of penile cancer (PeCa). Given its unique lymphatic spread patterns—where distant metastases rarely manifest without preceding involvement of the inguinal LNs—PeCa has subsequently been recognized as a promising candidate for extended SLNB investigation [3]. As a result, the EAU-ASCO Penile Cancer Guidelines recommend SLNB for T1b or higher PeCa patients [4].

Like in PeCa, SLNB's application in melanoma and breast cancer has shown promising results in clinical practice. In 1992, Morton et al. introduced SLNB for early-stage melanoma, later validating its accuracy in predicting regional node metastasis and establishing its prognostic significance for primary skin melanoma, stage T2–T3 [5, 6]. Similarly, in 1993, Krag et al. utilized radiolabeled colloid to identify sentinel nodes in breast cancer, marking a transformative shift in breast cancer management where sentinel lymph node biopsy (SLNB) gradually replaced routine axillary lymph node dissection due to its lower morbidity [7, 8].

Encouraging evidence for the potential use of the sentinel lymph node concept in the treatment of PeCa, melanoma, and breast cancer has led to intensive research in other areas of oncological surgery. In gynecologic cancers such as endometrial, cervical, and vulvar, emerging research suggests the potential of sentinel lymph node staging as an alternative to traditional lymphadenectomy [9–13]. SLNB shows promise for early-stage oropharyngeal squamous cell carcinoma for head and neck cancers but is less established for melanoma due to complex lymphatic drainage patterns [14, 15]. The role of SLNB in other cancers, including gastric, colorectal, and lung, is still being assessed, but preliminary findings suggest potential clinical benefits [16–18].

In urological malignancies such as kidney and bladder cancers, the significance of SLNB remains ambiguous. The EAU-ASCO Guidelines underscore SLNB as a promising avenue, particularly in the imaging context for muscle-invasive bladder cancer [2, 19]. For prostate cancer (PCa), SLNB is referenced in the guidelines as a research-focused staging method, but it currently lacks robust evidence confirming its efficacy [20].

As with many cancers, the decision to undertake lymph node dissection in the treatment of intermediate and high-risk patients is often met with uncertainty, and this holds true for PCa as well. Despite the widespread use of extended pelvic lymph node dissection (ePLND), a significant proportion of patients undergo this procedure unnecessarily, as evidenced by the absence of metastatic lymph nodes in the final pathology reports [21]. In this context, sentinel lymph node (SLN) dissection could serve as a valuable screening tool to distinguish patients who would benefit from ePLND from those who could safely avoid it. While SLNB in PCa shows promise, the evidence supporting its use is not as robust as for penile, breast cancer, or melanoma. More clinical data need to be collected to solidify its role in prostate cancer management.

The study aims to evaluate the technical details and efficacy of sentinel lymph node (SLN) dissection during laparoscopic radical prostatectomy in patients with clinically localized PCa, using indocyanine green/near-infrared (ICG/NIR) technology.

2. Material and methods

Between 2020 and 2023, a prospective study was conducted on 60 patients diagnosed with PCa, as confirmed by prostate biopsy, contrast-enhanced computerized tomography (CT) scans, bone scans. Verification of the affected areas of prostate was conducted using multiparametric magnetic resonance imaging (mpMRI) of the prostate gland prior to biopsy. Inclusion criteria for the study were clinically diagnosed and histologically confirmed localized prostate cancer, subjected to radical prostatectomy (**Table 1**). Patients with imaging confirmed lymphadenopathy, history of prostatitis, transurethral resection of prostate (TURP) due to benign prostatic hyperplasia (BPH), previous hormonal treatment were excluded from the study. Written informed consent was obtained from all patients in accordance with the Declaration of Helsinki. The local ethics committee was informed of this investigation.

All patients underwent laparoscopic radical prostatectomy supplemented with SLNs dissection, followed by standard ePLND in 45 patients with intermediate and high-risk PCa. The remaining 15 patients, who had clinically localized low-risk PCa, underwent only sentinel lymph node dissection.

To achieve ICG/NIR technology, we used Verdye (25 mg) Indocyanine green, Diagnostic Green, IMAGE1 S™ 4 K Rubina™ KARL STORZ equipment. The

Index	Overall (n = 60)	pN0 (n = 47)	pN1 (n = 13)
Age at diagnosis, years, median (IQR)	65.0 (60.75;69.0)	67.0 (62.0;69.0)	62 (59.0;65.0)
PSA, ng/ml, median (IQR)	12.05 (8.04;19.25)	10.6 (7.42;16.36)	15.33 (11.7;32.0)
BMI, kg/m^2 median (IQR)	27.34 (25.43;29.56)	26.89 (25.26; 29.26)	28.73 (26.12; 30.89)
Prostate volume, cm^3, median (IQR)	41.0 (34.0; 57.4)	38.0 (33.0;57.0)	52.0(40.0;57.0)
Risk group (No. pts., %):			
Low	15 (25%)	15 (32%)	0 (0%)
Intermediate	29 (49%)	25 (52%)	5 (38%)
High	16 (26%)	8 (16%)	8 (62%)
No. removed SLNs/ Mean ± SD			
Overall	160	151	39
Left side	81/1.3 ± 0.8	63/1.3 ± 0.7	18/1.2 ± 0.6
Right side	79/1.3 ± 0.7	58/1.3 ± 1.0	21/1.6 ± 0.6
No. removed LNs/ Mean ± SD			
Left side	223/9.7 ± 2.6	149/4.7 ± 2.7	74/4.9 ± 1.9
Right side	243/10.6 ± 2.3	158/5.7 ± 2.3	85/6.5 ± 2.4
%ISUP grade(No. pts):			
1	9 (15%)	9 (19%)	-
2	28 (47%)	24 (51%)	4 (31%)
3	10 (17%)	6 (13%)	4 (31%)
4–5	13 (21%)	8 (17%)	5 (38%)
% T stage (No. pts):			
2	18 (30%)	13 (34%)	6 (55%)
3a	15 (25%)	22 (56%)	4 (36%)
3b	27 (45%)	4 (10%)	1 (9%)

Table 1.
Patient parameters and pathological outcomes.

procedure of intraprostatic ICG injection was performed in the operating room just before the surgery began.

The patient was positioned in the lithotomy position. ICG was reconstituted immediately prior to use. To prepare the solution, 10 ml of sterile water was injected into the vial containing 25 mg of ICG powder, resulting in a concentration of 2.5 mg/ml. The vial was gently swirled until the ICG was completely dissolved, which typically takes 3–4 min. After the induction of general anesthesia, intraprostatic injection of indocyanine green (ICG) solution 2.5 mg/mL was performed using a Chiba needle 22 g under the guidance of a transrectal ultrasound, utilizing MRI-guided cognitive fusion of the lesions.

ICG solution was injected into the area of the MRI-identified tumor and sextant biopsy areas of the prostate (2.5 mL per lobe). Immediately following the injection, the patient was repositioned into a standard supine position for the radical prostatectomy. The positioning of the patient and trocar set-up were standard for a transperitoneal laparoscopic prostatectomy. Using the 25 to 30-degree Trendelenburg position, intestinal loops are moved cranially to expose the field of the common iliac vessels where lymph node luminescence is supposed to be determined [22]. An incision of the peritoneum along the iliac vessels is performed. After excision of fluorescent lymph nodes on the right and left side, the ePLND was performed in the patient with indications, followed by radical prostatectomy. All surgical procedures were performed by the same expert senior surgeon.

Patient demographic data, body mass index (BMI), pre-operative prostate-specific antigen (PSA) were collected for all participants. Decision for radical prostatectomy and ePLND was based on PSA, MRI, CT, and bone scan data. Tumor characteristics and staging were assessed through routine postoperative histopathological examination.

2.1 Data analysis

Descriptive statistics were calculated for all clinical characteristics. Continuous variables were reported as medians (interquartile range), and categorical variables as frequencies (percentage). Multiple linear regression was employed to model the relationship between characteristics of interest. Two-sided p-values were reported for all statistical tests. The likelihood of differences in categorical data was assessed using Pearson's Chi-square test ($\chi 2$), while quantitative and ordinal data were evaluated using the Mann-Whitney U test. The critical value for the level of statistical significance (p) was set at less than 5% ($p < 0.05$) for all types of analyses. Statistical analysis was performed using R version 3.5.1.

3. Results

3.1 Identification of sentinel lymph nodes

Upon completion of the ICG injection, patients were positioned in a standard supine position, and the surgical field and instruments were subsequently prepared. The incision and insertion of the first trocar took place, on average, 8–10 minutes following the ICG injection. After inserting the trocars and instruments, the abdominal cavity was examined using a near-infrared filter (**Figure 1**).

It was observed that the fluorescence of the lymph nodes was most commonly situated in the area of the bifurcation of the common iliac artery. This could be clearly seen after cranial displacement of the iliac and sigmoid colon loops. A cut is made

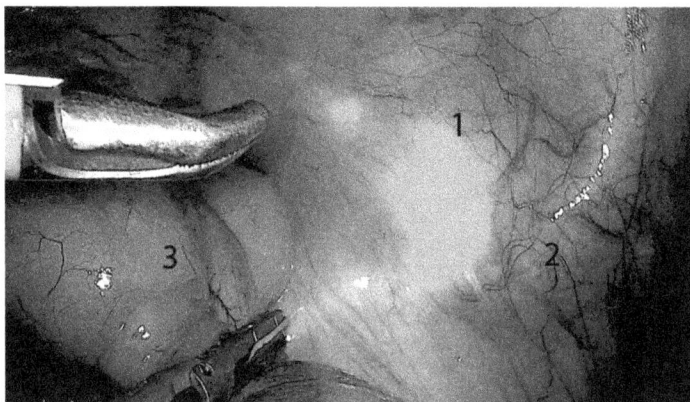

Figure 1.
Visualization of lymph node luminescence. 1 – Fluorescent lymph node, 2 – Right external iliac artery,
3 – Intestinal loop.

in the peritoneum along the iliac vessels, followed by meticulous "en bloc" removal of the fluorescence lymph nodes. To inhibit further ICG dissemination within the lymphatic system, surgical intervention commences at the cranial boundary of the dissection area and proceeds in a caudal direction. Adjacent lymphatic channels surrounding the node are sealed off, either through bipolar coagulation or using an ultrasonic scalpel (**Figure 2**). The dissection is finalized while avoiding direct handling or damage to the node tissue. Employing this approach mitigates the risk of ICG leakage, thereby preventing indiscriminate luminescence in the nearby tissue (**Figure 3**).

In 54 cases (90%) and 57 cases (95%), SLN fluorescence was identified in the area of the iliac vessels' bifurcation, with minor variations of 1–1.5 cm in the medial and lateral directions for the right and left sides, respectively. In 4 cases (7%) for the right side and 2 cases (3%) for the left side, the SLNs were localized in the obturator fossa. In one case for both sides, the SLNs fluorescence was found in the area near the upper vesical artery and vein of the internal iliac vessels. In one case, fluorescence was not identified on the right side.

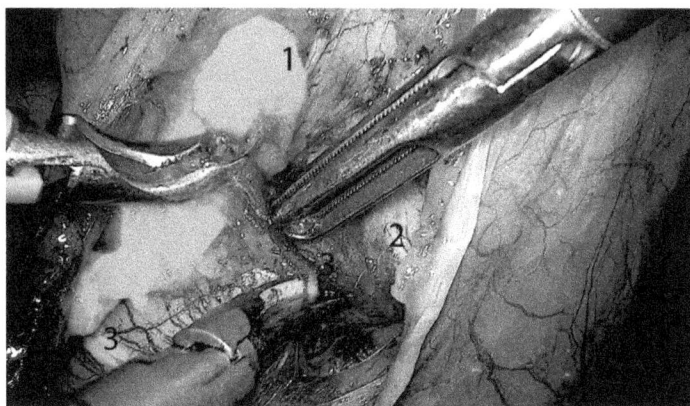

Figure 2.
Dissection of ICG+ lymph node. 1 – Fluorescent lymph node, 2 – Right external iliac artery, 3 – Lymphatic vessels.

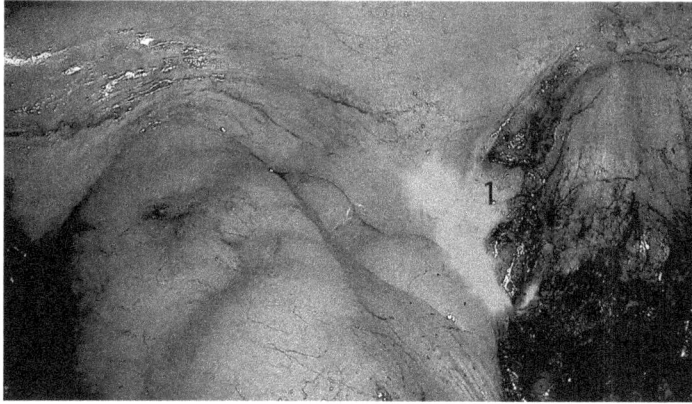

Figure 3.
Dissection of fluorescent lymph node is finished. 1 – ICG leakage in surrounding tissue.

3.2 Time characteristics for lymphadenectomy in prostate cancer patients

The time characteristics for lymphadenectomy in patients with prostate cancer are outlined in **Table 2**. In all cases, lymphadenectomy was initiated on the right side and then transitioned to the left. On average, lymphadenectomy on the right side commenced 14.5 (8.0; 21) minutes after the start of the surgery, and on the left side, it commenced after 40.0 (31.0; 52.0) minutes. The mean time to locate sentinel nodes on the right side was 27.5 (23.5; 30.0) minutes; on the left, it was 46.5 (34.5; 52.5) minutes. The total duration for node removal based on extended lymph node dissection was 22.0 (15.0; 35.0) minutes on the right and 15.5 (13.0; 23.0) minutes on the left. The time specifically for SLNs was 6.0 (3.0; 9.0) minutes on the right and 4.0 (2.0; 6.0) minutes on the left. The minimum duration for sentinel node excision was 1 minute, and the maximum was 18 minutes.

The difference in duration for this stage of the operation based on the side of the node location was not statistically significant ($p > 0.05$), although there was a trend toward increased time for procedures on the right side due to anatomical complexities and concomitant pathological processes (adhesive process) in operated patients.

Indicators	Node location	Right side	Left side	p-value
Time of initiation, min	Sentinel lymph nodes	14.5 (8.0; 21.0)	40.0 (31.0; 52.0)	NA
	Other lymph nodes	27.5 (23.5; 30.0)	46.5 (34.5; 52.5)	NA
Duration of lymph node removal procedure, min	Sentinel lymph nodes	6.0 (3.0; 9.0)	4.0 (2.0; 6.0)	0.161
	Other lymph nodes	14.0 (11.0; 25.0)	12.0 (9.0; 17.0)	0.300
Total, min		22.0 (15.0; 35.0)	15.5 (13.0; 23.0)	0.259

Table 2.
Average characteristics of time taken for lymphadenectomy in patients with prostate cancer, Me (25%; 75%).

It was noted that the rate of SLN detection depended on several factors. The most significant factor was the presence of excess adipose tissue. In this patient group, initial visualization of the SLNs was not possible. These lymph nodes could only be identified after incision of the parietal peritoneum and dissection of the surrounding adipose tissue. When sentinel nodes were located in atypical areas (not in the area of the bifurcation of the common iliac artery), the time required for their identification increased.

We evaluate the impact of BMI on the process of localizing SLNs. It was found that the time taken to locate these nodes directly correlates with BMI: right side (r = 0.49, p = 0.029); left side (r = 0.47, p = 0.035) (**Figure 4**).

The average time to locate sentinel nodes on the right side from the start of the operation was 27.5 (23.5; 30.0) minutes. Specifically, with a normal BMI (up to 25 kg/m^2), the time is 26.0 (16.0; 28.0); for overweight BMI (up to 30 kg/m^2) - 29.0 (23.0; 33.0); and for obesity stage I (up to 35 kg/m^2) - 29.0 (24.0; 39.0) minutes. The average time to locate nodes on the left side was 46.5 (32.5; 52.5) minutes. In terms of BMI: normal - 35.0 (26.0; 51.0), overweight - 47.0 (34.0; 52.0), and obesity stage I -54.0 (40.0; 69.0) minutes (**Table 3**).

Unquestionably, BMI is a factor that technically complicates both open and laparoscopic surgical interventions. However, the influence of BMI on the process of SLN removal was statistically significant based on the collected data. Excessive body weight technically complicates and likely increases the time required to perform ePLND. Adhesion formation in the area adjacent to the aortic vessels led to the need for adhesiolysis, which also increased the time to visualize sentinel nodes. However, this factor is less significant than obesity, as it was infrequent—in 7 (11%) patients.

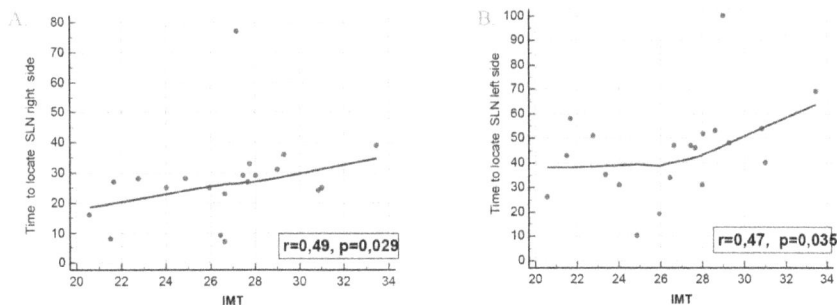

Figure 4.
Correlation between time to identify SLNs on the right (A) and left (B) sides and the patient's BMI.

BMI, kg/m^2	Time, min	
	Right side	**Left side**
Normal (18.5–24.9)	26.0 (16.0; 28.0)	35.0 (26.0; 51.0)
Overweight (25–29.9)	29.0 (23.0; 33.0)*	47.0 (34.0; 52.0)*
Obesity Stage I (30–34.9)	29.0 (24.0; 39.0)*	54.0 (40.0; 69.0)*

$p < 0.05$ compared to the normal range.

Table 3.
Average time to locate SLNs from the start of surgery based on BMI, Me (25%; 75%).

3.3 Pathohistological assessment

Identification and removal of SLNs were conducted in 60 patients. In 45 patients categorized as intermediate and high-risk according to the EAU risk group classification, extended lymphadenectomy was performed. A total of 160 SLNs were identified in these 60 patients, with 81 located on the left side and 79 on the right. In the 45 patients who underwent extended lymphadenectomy, 466 lymph nodes were removed: 223 on the left side and 243 on the right, with averages of 9.69 ± 2.61 and 10.56 ± 2.3, respectively.

Regional lymph node metastases were observed among the patients studied in 12 cases (12%), predominantly in those with pT3 stage disease. On the left side, the number of identified SLNs ranged from 0 to 3, averaging 1.35 ± 0.78. On the right, the range was from 0 (in 3 cases) to 3, with an average of 1.32 ± 0.65. There was no statistical difference between the two sides (p = 0.683).

Histological examination revealed no metastatic lymph node involvement in patients from the low-risk group where only SLNs were removed. In the remaining 45 patients, SLN involvement was found in 12 cases (22%) or in 17 out of the 160 removed SLNs (11%). Of these, 10 cases showed unilateral SLN involvement, while 2 exhibited bilateral involvement. In 4 of these 12 specific cases, isolated SLN involvement was observed, whereas in 8 cases, metastatic involvement had spread to other lymph nodes removed during extended lymphadenectomy. In one patient, a non-SLN showed involvement, but SLN involvement was not observed.

To evaluate the predictive capability of sentinel metastases for anticipating regional lymph node metastases, a chi-square test was conducted. The test revealed a specificity (Sp) of 90%, sensitivity (Se) of 80%, positive predictive value (PPV) of 61.5%, and negative predictive value (NPV) of 95.7%.

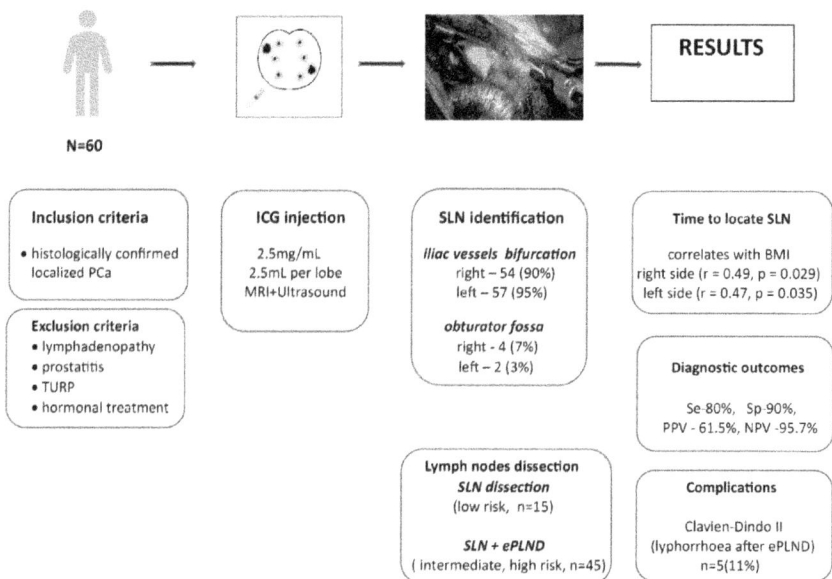

Figure 5.
Study design and obtained results.

In our study, 5 patients (11%) experienced complications categorized as Clavien-Dindo Grade II. Specifically, these patients exhibited prolonged lymphorrhoea, necessitating an extended drainage duration without the need for additional procedures. Notably, these complications were observed exclusively in patients who underwent ePLND. Given these findings, the potential to reduce ePLND-associated complications by assessing SLN presents an appealing surgical option for patients with prostate cancer.

A concise overview of the study design and the results obtained is illustrated in **Figure 5**.

4. Discussion

Our study focuses on exploring the capabilities of SLN biopsy exclusively using ICG technology. Currently, many studies for SLN verification employ magnetic and radioactive markers, each with its advantages and disadvantages, as well as their combined use with ICG [23]. Jeschke S. et al. found that ICG performed as well as 99mTc, and combining the two substances did not yield significantly better results than ICG alone. Utilizing the outlined fluorescence navigation method allows for real-time visualization of both lymph nodes and lymphatic vessels. This technique appears to offer comparable effectiveness to SLN dissection while being simpler to implement [22].

The procedure for injecting ICG into the prostate gland to identify SLNs is not standardized. Various methods for ICG injection have been described, including transrectal, transperineal, percutaneous intraoperative, and cystoscopic approaches. A study by Manny et al. compared these methods and found that robot-guided percutaneous ICG injection into the prostate was the most cost- and time-effective option [24]. Previously noted limitations of ultrasound in visualizing prostate lesions [25–27] have been overcome by combining it with MRI data. The use of MRI capability to visualize the zonal anatomy of the prostate and tumor foci, as described by the Prostate Imaging Reporting and Data System (PIRADS) version 2, has led to the fusion of MR images with real-time transrectal ultrasound to improve biopsy accuracy in diagnosing prostate cancer [28]. This technique has provided a more accurate solution for tracer injection into the affected zone of the prostate, thereby reducing the value of percutaneous ICG injection, which does not offer this capability. According to the FUTURE study, the existing three techniques—MRI-transrectal ultrasound (TRUS) fusion, cognitive registration, or in-bore MRI—have roughly equivalent diagnostic capabilities [29]. For preoperative use, the most cost- and time-effective option is the cognitive method, which we employed in our study.

In our study, we successfully employed transrectal injection of ICG, which was not accompanied by adverse side effects. By utilizing exclusion criteria such as a history of chronic prostatitis, which leads to prostate gland sclerosis, and prior transurethral resection of the prostate (TURP), the likelihood of ICG extravasation and inadequate distribution within the prostate tissue was reduced. Transrectal and transperineal tracer injections are currently being used [30, 31]. However, the preference is increasingly leaning toward the transperineal method under ultrasound guidance. This is because the needle is introduced parallel to the axis of the peripheral zone, allowing for the complete application of the tracer within this area without risk of extravasation. Additionally, this approach substantially reduces the risk of infection [32].

In the scientific literature, various methods of tracer injection into the prostate gland are described, primarily targeting the peripheral zone and focusing on the basal and apical portions, including the transitional zone. Buckle T et al. found that injecting ICG into the peripheral zone led to the visualization of a greater number of SLNs than injecting into the prostate's central zone. This suggests that the location of intraprostatic tracer deposition may influence the preoperatively visualized lymphatic drainage [27].

Given the variability in tracer deposition, there is ongoing debate about whether to employ intraprostatic or intralesional tracer injections. Prostate cancer is generally a multifocal disease, predominantly originating in the peripheral zone. A recent randomized Phase II trial by Wit et al. revealed that intralesional injections were more effective in detecting lymph node metastases in the final pathology report. However, this technique failed to identify positive lymph nodes arising from diffuse or non-index prostatic lesions. Therefore, it is concluded that an optimal tracer administration strategy would likely involve a combination of both intraprostatic and intralesional approaches [33].

According to published data, the volume of ICG injected into the prostate gland varies, ranging from 0.1 to 1 mL per injection site, with concentrations varying from 0.1 to 1 mg/mL [27, 30, 32, 34]. There is currently no information favoring one approach over the others. From this perspective, reducing the concentration of the agent and increasing its volume aligns with the concept that the agent should be introduced not only into the areas of lesions identified by MRI but also into other parts of the prostate. Consequently, we chose the method proposed by Hruby et al., which involves the injection of a 0.1 mg/mL ICG solution at a volume of 2.5 mL per lobe, divided into 3 locations [32].

The time parameters of ICG distribution within the lymphatic drainage of the prostate gland remain poorly defined. According to published literature, ICG disperses relatively quickly—within 5 to 30 minutes—through the prostate's lymphatic system due to the small molecular size of ICG [24, 32, 35, 36]. Therefore, precise identification of SLNs appears to be challenging. We monitored the timing of lymph node detection and excision, displaying luminescence features by sequentially performing the procedure first on the left side and then on the right. The time for the completion of the procedure varied within one hour from the moment of ICG injection to the start of surgical intervention. In this context, we found no significant differences in either the localization of the removed lymph nodes or their numbers, which were largely consistent with data reported by other authors [22, 36, 37].

The most significant factor influencing the speed of detection and removal of SLNs was the presence of obesity. This aligns with data from Stoffels et al., who demonstrated in a prospective study comparing ICG to 99mTc in melanoma patients that free-ICG has a tissue penetration depth of only 10–15 mm. This limits its ability to detect lymphatic structures that extend into deeper tissues, a limitation attributed to ICG's inherently low fluorescence and the strong attenuation of low-energy NIRF in tissue [38].

The data we obtained regarding the predictive features of SLNB align with the broader picture presented by other authors for ICG technology and its combination with 99mTc-nanocolloid, as analyzed by Rossin et al. The "per patient" Se ranged from 75 to 100%, NPV from 93.8–100%, with false negatives (FN) at 0.0%, and false positives (FP) at 0.0%. On a "per node" basis, Se ranged from 34.1–100%, Sp from 64.8–99.0%, and NPV from 98.1–98.2% [23]. It is worth noting that these metrics could be influenced by the method of tracer injection into the prostate gland.

Time parameters can also impact the identification of sentinel nodes. The longer the time elapsed since the injection of ICG, the greater the likelihood of its spreading to lymph nodes of the second and third tiers. While this allows for better identification of the lymphatic drainage pathways from the prostate, it decreases the likelihood of accurately identifying SLNs. Based on ICG lymphography data, five main potential lymphatic pathways and sites can be distinguished: an internal route, a lateral route, a presacral route, a paravesical artery site, and a pre-prostatic site [30, 39].

Hruby et al. showed that the excision of ICG-positive lymph node groups resulted in increased Se (97.7%), NPV (99%), and accuracy (71%), while decreasing the false-negative rate (2%) when compared to extended and super-extended PLND. The use of fluorescence-guided pelvic lymph node dissection enhances the reliable identification of the prostate's lymphatic drainage. Concentrating solely on nodes directly draining the prostate reduces the number of nodes excised, yet diagnostic precision is elevated. Additionally, this technique obviates the need to distinguish between initial receiving sites (sentinel nodes) and subsequent tiers, a task that can be challenging but is crucial in traditional sentinel node concepts [32].

5. Conclusions

The study suggests that SLN dissection is an effective, safe, and feasible approach for preoperative clinical nodal staging in prostate cancer patients, offering a viable approach for those with intermediate and high-risk PCa. Factors such as BMI were found to influence the time taken to locate the SLNs, which can affect the surgical process. The issue of accurately identifying SLNs for the prostate remains controversial and requires further research to clarify anatomical and temporal parameters of lymph node dissection.

Conflict of interest

The authors declare that they have no conflicts of interest.
This research has received no external funding.

Author details

Robert M. Molchanov[1]*, Oleg B. Blyuss[2] and Ruslan V. Duka[1]

1 Dnipro State Medical University, Dnipro, Ukraine

2 Queen Mary University of London, London, United Kingdom

*Address all correspondence to: rob_molch@yahoo.com

IntechOpen

References

[1] Singh P, Kaul P, Singhal T, Kumar A, Garg PK, Narayan ML. Role of sentinel lymph node drainage mapping for localization of contralateral lymph node metastasis in locally advanced oral squamous cell carcinoma – A prospective pilot study. Indian Journal of Nuclear Medicine. 2023;**38**(2):125-133. DOI: 10.4103/ijnm.ijnm_120_22

[2] Malkiewicz B, Kielb P, Kobylanski M, Karwacki J, Poterek A, Krajewski W, et al. Sentinel lymph node techniques in urologic oncology: Current knowledge and application. Cancers. 2023;**15**(9):2495. DOI: 10.3390/cancers15092495

[3] Cabanas RM. An approach for the treatment of penile carcinoma. Cancer. 1977;**39**(2):456-466. DOI: 10.1002/1097-0142(197702)

[4] Brouwer OR, Albersen M, Parnham A, Protzel C, Pettaway CA, Ayres B, et al. European Association of Urology-American Society of Clinical Oncology Collaborative Guideline on Penile Cancer: 2023 Update. European Urology. 2023;**83**(6):548-560. DOI: 10.1016/j.eururo.2023.02.027

[5] Morton DL, Thompson JF, Cochran AJ, Mozzillo N, Nieweg OE, Roses DF, et al. Final trial report of sentinel-node biopsy versus nodal observation in melanoma. The New England Journal of Medicine. 2014;**370**(7):599-609. DOI: 10.1056/NEJMoa1310460

[6] Holmberg CJ, Mikiver R, Isaksson K, Ingvar C, Moncrieff M, Nielsen K, et al. Prognostic significance of sentinel lymph node status in thick primary melanomas (> 4 mm). Annals of Surgical Oncology. 14 Aug 2023. DOI: 10.1245/s10434-023-14050-w. PMID: 37574516. Available from: https:// pubmed.ncbi.nlm.nih.gov/37574516/ [Epub ahead of print]

[7] Krag DN, Weaver DL, Alex JC, Fairbank JT. Surgical resection and radiolocalization of the sentinel lymph node in breast cancer using a gamma probe. Surgical Oncology. 1993;**2**(6):335-339; Discussion 40. DOI: 10.1016/0960-7404(93)90064-6

[8] Krag DN, Anderson SJ, Julian TB, Brown AM, Harlow SP, Costantino JP, et al. Sentinel-lymph-node resection compared with conventional axillary-lymph-node dissection in clinically node-negative patients with breast cancer: Overall survival findings from the NSABP B-32 randomised phase 3 trial. The Lancet Oncology. 2010;**11**(10):927-933. DOI: 10.1016/S1470-2045(10)70207-2

[9] Capozzi VA, Rosati A, Maglietta G, Vargiu V, Scarpelli E, Cosentino F, et al. Long-term survival outcomes in high-risk endometrial cancer patients undergoing sentinel lymph node biopsy alone versus lymphadenectomy. International Journal of Gynecological Cancer. 2023;**33**(7):1013-1020. DOI: 10.1136/ijgc-2023-004314

[10] Obermair A. Sentinel lymph node staging versus lymphadenectomy in high risk patients: Is there sufficient evidence to change practice? International Journal of Gynecological Cancer. 2023;**33**(7):1021-1022. DOI: 10.1136/ijgc-2023-004662

[11] Matsuo K, Klar M, Barakzai SK, Jooya ND, Nusbaum DJ, Shimada M, et al. Utilization of sentinel lymph node biopsy in the early ovarian cancer surgery. Archives of Gynecology and Obstetrics. 2023;**307**(2):525-532. DOI: 10.1007/s00404-022-06595-0

[12] Yahata H, Kobayashi H, Sonoda K, Kodama K, Yagi H, Yasunaga M, et al. Prognostic outcome and complications of sentinel lymph node navigation surgery for early-stage cervical cancer. International Journal of Clinical Oncology. 2018;**23**(6):1167-1172. DOI: 10.1007/s10147-018-1327-y

[13] Matsuo K, Chen L, Robison K, Klar M, Roman LD, Wright JD. Trends in the use of indocyanine green for sentinel lymph node mapping in vulvar cancer. American Journal of Obstetrics and Gynecology. 2023;**229**(4):466-468. DOI: 10.1016/j.ajog.2023.07.019

[14] Gupta T, Maheshwari G, Kannan S, Nair S, Chaturvedi P, Agarwal JP. Systematic review and meta-analysis of randomized controlled trials comparing elective neck dissection versus sentinel lymph node biopsy in early-stage clinically node-negative oral and/or oropharyngeal squamous cell carcinoma: Evidence-base for practice and implications for research. Oral Oncology. 2022;**124**:105642. DOI: 10.1016/j.oraloncology.2021.105642

[15] Pasha T, Arain Z, Buscombe J, Aloj L, Durrani A, Patel A, et al. Association of complex lymphatic drainage in head and neck cutaneous melanoma with sentinel lymph node biopsy outcomes: A Cohort Study and Literature Review. JAMA Otolaryngology. Head & Neck Surgery. 2023;**149**(5):416-423. DOI: 10.1001/jamaoto.2023.0076

[16] Huang Y, Pan M, Deng Z, Ji Y, Chen B. How useful is sentinel lymph node biopsy for the status of lymph node metastasis in cT1N0M0 gastric cancer? A systematic review and meta-analysis. Updates in Surgery. 2021;**73**(4):1275-1284. DOI: 10.1007/s13304-021-01026-2

[17] Di Berardino S, Capolupo GT, Caricato C, Caricato M. Sentinel lymph

node mapping procedure in T1 colorectal cancer: A systematic review of published studies. Medicine. 2019;**98**(28):e16310. DOI: 10.1097/MD.0000000000016310

[18] Konno H, Minamiya Y. Sentinel lymph node diagnosis for lung cancer. Kyobu Geka. 2018;**71**(10):875-879

[19] Witjes JA, Bruins HM, Cathomas R, Comperat EM, Cowan NC, Gakis G, et al. European Association of Urology Guidelines on Muscle-invasive and Metastatic Bladder Cancer: Summary of the 2020 Guidelines. European Urology. 2021;**79**(1):82-104. DOI: 10.1016/j.eururo.2020.03.055

[20] Mottet N, van den Bergh RCN, Briers E, Van den Broeck T, Cumberbatch MG, De Santis M, et al. EAU-EANM-ESTRO-ESUR-SIOG guidelines on prostate cancer-2020 Update. Part 1: Screening, diagnosis, and local treatment with curative intent. European Urology. 2021;**79**(2):243-262. DOI: 10.1016/j.eururo.2020.09.042

[21] Gandaglia G, Ploussard G, Valerio M, Mattei A, Fiori C, Fossati N, et al. A novel nomogram to identify candidates for extended pelvic lymph node dissection among patients with clinically localized prostate cancer diagnosed with magnetic resonance imaging-targeted and systematic biopsies. European Urology. 2019;**75**(3):506-514. DOI: 10.1016/j.eururo.2018.10.012

[22] Jeschke S, Lusuardi L, Myatt A, Hruby S, Pirich C, Janetschek G. Visualisation of the lymph node pathway in real time by laparoscopic radioisotope- and fluorescence-guided sentinel lymph node dissection in prostate cancer staging. Urology. 2012;**80**(5):1080-1086. DOI: 10.1016/j.urology.2012.05.050

[23] Rossin G, Zorzi F, De Pablos-Rodriguez P, Biasatti A, Marenco J,

Ongaro L, et al. Sentinel lymph node biopsy in prostate cancer: An overview of diagnostic performance, oncological outcomes, safety, and feasibility. Diagnostics (Basel). 2023;**13**(15):2543. DOI: 10.3390/diagnostics13152543

[24] Manny TB, Patel M, Hemal AK. Fluorescence-enhanced robotic radical prostatectomy using real-time lymphangiography and tissue marking with percutaneous injection of unconjugated indocyanine green: The initial clinical experience in 50 patients. European Urology. 2014;**65**(6):1162-1168. DOI: 10.1016/j.eururo.2013.11.017

[25] Holl G, Dorn R, Wengenmair H, Weckermann D, Sciuk J. Validation of sentinel lymph node dissection in prostate cancer: Experience in more than 2,000 patients. European Journal of Nuclear Medicine and Molecular Imaging. 2009;**36**(9):1377-1382. DOI: 10.1007/s00259-009-1157-2

[26] Joniau S, Van den Bergh L, Lerut E, Deroose CM, Haustermans K, Oyen R, et al. Mapping of pelvic lymph node metastases in prostate cancer. European Urology. 2013;**63**(3):450-458. DOI: 10.1016/j.eururo.2012.06.057

[27] Buckle T, Brouwer OR, Valdes Olmos RA, van der Poel HG, van Leeuwen FW. Relationship between intraprostatic tracer deposits and sentinel lymph node mapping in prostate cancer patients. Journal of Nuclear Medicine. 2012;**53**(7):1026-1033. DOI: 10.2967/jnumed.111.098517

[28] Drost FH, Osses D, Nieboer D, Bangma CH, Steyerberg EW, Roobol MJ, et al. Prostate magnetic resonance imaging, with or without magnetic resonance imaging-targeted biopsy, and systematic biopsy for detecting prostate cancer: A cochrane systematic review and meta-analysis. European Urology.

2020;**77**(1):78-94. DOI: 10.1016/j.eururo.2019.06.023

[29] Wegelin O, Exterkate L, van der Leest M, Kummer JA, Vreuls W, de Bruin PC, et al. The FUTURE Trial: A multicenter randomised controlled trial on target biopsy techniques based on magnetic resonance imaging in the diagnosis of prostate cancer in patients with prior negative biopsies. European Urology. 2019;**75**(4):582-590. DOI: 10.1016/j.eururo.2018.11.040

[30] Shimbo M, Endo F, Matsushita K, Hattori K. Impact of indocyanine green-guided extended pelvic lymph node dissection during robot-assisted radical prostatectomy. International Journal of Urology: Official Journal of the Japanese Urological Association. 2020;**27**(10):845-850. DOI: 10.1111/iju.14306

[31] Polom K, Murawa D, Polom W, et al. Intraoperative laparoscopic fluorescence guidance to the sentinel lymph node in prostate cancer patients: Clinical proof of concept of an integrated functional imaging approach using a multimodal tracer. European Urology. 2012;**61**(3):e18. DOI: 10.1016/j.eururo.2011.12.002

[32] Hruby S, Englberger C, Lusuardi L, Schatz T, Kunit T, Abdel-Aal AM, et al. Fluorescence guided targeted pelvic lymph node dissection for intermediate and high risk prostate cancer. The Journal of urology. 2015;**194**(2):357-363. DOI: 10.1016/j.juro.2015.03.127

[33] Wit EMK, van Beurden F, Kleinjan GH, Grivas N, de Korne CM, Buckle T, et al. The impact of drainage pathways on the detection of nodal metastases in prostate cancer: A phase II randomized comparison of intratumoral vs intraprostatic tracer injection for sentinel node detection. European Journal of Nuclear Medicine and Molecular Imaging. 2022;**49**(5):1743-1753. DOI: 10.1007/s00259-021-05580-0

[34] Claps F, Ramirez-Backhaus M, Mir Maresma MC, Gomez-Ferrer A, Mascaros JM, Marenco J, et al. Indocyanine green guidance improves the efficiency of extended pelvic lymph node dissection during laparoscopic radical prostatectomy. International Journal of Urology: Official Journal of the Japanese Urological Association. 2021;**28**(5):566-572. DOI: 10.1111/iju.14513

[35] Inoue S, Shiina H, Arichi N, Mitsui Y, Hiraoka T, Wake K, et al. Identification of lymphatic pathway involved in the spreading of prostate cancer by fluorescence navigation approach with intraoperatively injected indocyanine green. Canadian Urological Association journal = Journal de l'Association des urologues du Canada. 2011;**5**(4):254-259. DOI: 10.5489/cuaj.10159

[36] Nguyen DP, Huber PM, Metzger TA, Genitsch V, Schudel HH, Thalmann GN. A specific mapping study using fluorescence sentinel lymph node detection in patients with intermediate- and high-risk prostate cancer undergoing extended pelvic lymph node dissection. European Urology. 2016;**70**(5):734-737. DOI: 10.1016/j.eururo.2016.01.034

[37] Mazzone E, Dell'Oglio P, Grivas N, Wit E, Donswijk M, Briganti A, et al. Diagnostic value, oncologic outcomes, and safety profile of image-guided surgery technologies during robot-assisted lymph node dissection with sentinel node biopsy for prostate cancer. Journal of Nuclear Medicine. 2021;**62**(10):1363-1371. DOI: 10.2967/jnumed.120.259788

[38] Stoffels I, Dissemond J, Poppel T, Schadendorf D, Klode J. Intraoperative fluorescence imaging for sentinel lymph node detection: Prospective clinical trial to compare the usefulness of indocyanine Green vs Technetium Tc 99m for identification of sentinel lymph nodes.

JAMA Surgery. 2015;**150**(7):617-623. DOI: 10.1001/jamasurg.2014.3502

[39] Yuen K, Yamashita M. Intraoperative fluorescence imaging using indocyanine green for detection of sentinel lymph nodes during prostatectomy. Nihon Rinsho. 2016;**74**(Suppl. 3):714-719

Chapter 6

Exploring the Role of the Lymphatic System in Immune Regulation: Implications for Autoimmunity, Cancer, and Infection

Marzieh Norouzian and Soghra Abdi

Abstract

The lymphatic system is the immune system's transport network (lymphatic vessels and lymph) that collects microbial antigens at the entrance and delivers them to the lymph nodes, where specific immune responses are stimulated. The lymphatic system maintains peripheral tolerance under normal conditions and rapidly develops protective immunity against foreign antigens after stimulation. Available evidence indicates that lymphatic function can be altered in various disease states such as cancer, infectious diseases, and autoimmunity. Many pathological conditions induce lymphangiogenesis, which is thought to provide an extensive lymphatic network that allows antigens and fluids to have greater access to the lymphatics. However, the role of lymphangiogenesis and lymphatic dysfunction in immune regulation is unclear. Understanding the causes of lymphatic dysfunction in pathological diseases will help develop new therapeutic approaches targeting the lymphatic system in various diseases. This chapter summarizes current knowledge about how lymphatic function is altered in autoimmune conditions, cancer, and infectious diseases, and how it modulates the immune response.

Keywords: lymphatic function, immune regulation, cancer, autoimmunity, infection

1. Introduction

One component of the circulatory system is the lymphatic system, which plays an important role in both immunological function and the drainage of excess extracellular fluid. The lymphatic system is also considered part of the circulatory and immune systems. The role of the lymphatic system is to regulate the body's fluid balance and filter pathogens from the blood, complementing the functions of the bloodstream. A complex network of lymphatic vessels connecting the local tissue site with secondary lymphatic organs such as lymph nodes, spleen, and mucosal lymphoid tissues constitutes the peripheral lymphatic system [1]. The peripheral lymphatic system is

IntechOpen

the main route for leukocyte transfer and antigen presentation. However, there is a broader view of how the lymphatic system influences immune responses beyond the physical connection between peripheral tissues and secondary lymphoid organs.

The lymphatic vessels, which provide structural and functional support for the distribution of antigens and antigen-presenting cells to drain lymph nodes, are known to be involved in an immune response. The lymphatic vasculature plays a critical role in maintaining peripheral tolerance or generating a protective immune response to infection or vaccination [2].

The immune system is regulated at several levels, both actively and passively. The active mechanism for regulating the immune response in the lymphatic system is to regulate the entry and migration of immune cells, expression of cytokines, chemokines, and adhesion molecules by lymphatic endothelial cells (LECs) [3]. LECs are specialized subsets of the endothelium of lymphatic vessels in tissues and lymph nodes that are essential for maintaining vascular integrity and proper lymphatic function. LECs in lymphatic vessels and lymph nodes provide a highly efficient pathway for initiating immune responses. LECs can recruit immune cells such as B cells, T cells, and dendritic cells (DCs) to the lymph nodes through the secretion of various chemokines and play a role in antigen presentation or exchange. Recruitment of immune cells is beneficial for the coordination of expansion and contraction of LECs and lymph nodes [3]. Recent studies have shown that cell surface molecules like PDL1 and interferon receptors are essential for the coordination of LEC division and death. During homeostasis, lymphatic endothelial cells play a role in immune regulation and tolerance induction through upregulation of PD-L1 expression and the absence of costimulatory molecules. In addition, after infection with or vaccination of viruses, LECs may be capable of obtaining, presenting, and exchanging foreign antigens [4]. However, in chronic disease, significant phenotypic changes in lymphatic endothelium due to transcription factor gene alterations have been reported [5]. There seems to be a need for more research into inflammation causing changes in lymphatic junctions, which are thought to influence immunological cell trafficking and the resolution of tissue inflammation.

Lymphoedema is the most obvious example of lymphatic and immune dysfunction interacting. During the progression of lymphoedema, the fluid build-up and congestion lead to several tissue changes, such as fibrosis and chronic inflammation. These changes, combined with the inability to deliver antigen and antigen-presenting cells to the lymph node, lead to a progressive decline in local immune function [6–8].

Under normal conditions, the lymphatic system is involved in controlling inflammatory responses and in maintaining tolerance. Not surprisingly, lymphatic dysfunction is associated with inflammation, cancer development and metastasis, infectious diseases, and sepsis, as the lymphatic system plays an important role in many physiological processes [9]. Many serious complications and abnormalities in the lymphatic system have recently been shown to be associated with changes in lymphatic transport function and lymphoid regulation of immune responses [6]. A wide range of diseases results in lymphangiogenesis, which can lead to an enlarged lymphatic network, allowing greater access for antigens and fluids into the lymphatic vessels. During inflammation and the progression of cancer, the growth of lymphatic vessels is often observed as lymphangiogenesis in the lymphatic capillaries and also in the lymph nodes. An enlarged lymphatic network appears to provide a greater surface area for fluid or cell entry into the lymphatic vessels [10, 11]. In any disease state, lymphatic function is also altered. However, the role of lymphangiogenesis and altered lymphatic function in regulating the immune system remains to be elucidated.

The authors focus on cancer, autoimmunity, and infectious diseases in the review due to the following motivations:

1. Disease prevalence: Cancer, autoimmunity, and infectious diseases are significant health concerns globally, affecting a large number of individuals.

2. Impact on lymphatic system: These diseases have been observed to induce alterations in lymphatic function, including lymphangiogenesis and lymphatic dysfunction.

3. Immune response modulation: Understanding how the lymphatic system is affected in these diseases can provide insights into how it modulates the immune response, which is crucial for developing effective therapeutic approaches.

4. Clinical relevance: By studying the lymphatic system's role in cancer, autoimmunity, and infectious diseases, researchers can potentially identify new targets for therapeutic interventions and improve patient outcomes.

Overall, the focus on these specific diseases allows for a comprehensive understanding of how lymphatic function is altered and its implications for immune regulation in various pathological conditions.

2. Immunological function of the lymphatic system in pathological conditions

The lymphatic system may be affected by a variety of conditions. Some occur during development or early childhood. Other diseases or injuries have caused the development of others [12]. Some of the most common problems in the lymphatic system include infections, blockage, and cancer [13]. The question of how lymphoid function or dysfunction contributes to the disruption of immunological homeostasis has been a long-standing topic in the field due to their diverse roles in regulating leukocyte trafficking and function. Here is a review of the immunological role of the lymphatic system in pathological conditions including cancer, autoimmunity, and infectious disease (**Figure 1**).

2.1 Cancer

The tumor microenvironment is a complex and dynamic network composed of cellular and noncellular components. Cancer-associated fibroblasts and infiltrating immune cells constitute tumor stroma and are critical components of the tumor microenvironment. Through various cytokines, chemokines, growth factors, and the release of extracellular matrix (ECM) proteins, the cellular components of the tumor microenvironment form a complex crosstalk with the tumor [14].

In many types of tumors, the inflamed microenvironment, in combination with biochemical factors such as low oxygen concentration, contributes to tumorigenesis, angiogenesis, and lymphangiogenesis. These factors cause immune cell dysregulation through different mechanisms including apoptosis of cytotoxic T cells and activation of suppressor T cells [15–17]. Cancer-induced lymphangiogenesis contributes to an expanded lymphatic network that increases the delivery of molecular and cellular

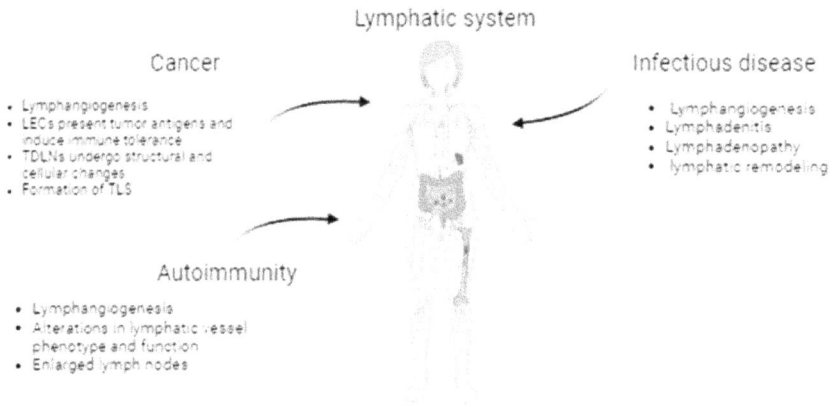

Figure 1.
A schematic diagram summarizing the effects of various pathological conditions on the lymphatic system. LECs: Lymphatic endothelial cells. TDLNs: Tumor-draining lymph nodes. TLS: Tertiary lymphoid structures.

components to the draining lymph node. Before tumor seeding, lymphangiogenesis of the lymph nodes (LNs) is established and supports the initial regional metastatic progression. Indeed, lymphangiogenesis has been correlated with lymph node involvement and a poor prognosis in patients with cancer [11].

Immune tolerance to primary tumors may be the cause of poor prognosis in patients with lymph node metastases. However, the nature of these effects is not well defined. Peripheral tolerance may occur due to chronic inflammation mediators caused by cancer, tumor proliferation factors, or tumor antigens from apoptotic tumor cells. In addition, lymphatic endothelial cells can present tumor antigens and induce immune tolerance. Several studies have reported the immunomodulatory effect of the lymphatic endothelium in the tumor microenvironment [18]. The release of a number of suppressive cytokines and chemokines and the expression of inhibitory molecules by lymphatic endothelium create an immunosuppressive microenvironment that leads to the inhibition of dendritic cell maturation and the induction of T-cell tolerance, further suppressing anti-tumor immunity and facilitating tumor escape from the immune system, followed by its growth and metastasis [19, 20]. It is well established that lymphatic endothelial cells are competent to present antigens and have the ability to induce tolerance of tumor-specific T cells. In human and murine models of melanoma and breast cancer, LECs have been shown to present tumor-associated antigens (TAAs) to CD4+ and CD8+ T cells and induce their suppression. LECs are also able to induce reduced CD86 expression on dendritic cells [21–23]. A study using melanoma and colon cancer cell lines has shown that IFN-γ expression in tumor-specific CD8+ T cells induces PD-L1 expression in lymphatic endothelial cells [19]. This leads to functional impairment of T cells in tumor cell lysis. These findings, together with the observation that LECs increase the suppressive function of regulatory T cells [20], lead to the conclusion that LECs play an immunosuppressive role in the breast cancer microenvironment.

As lymphangiogenesis or immune modulation by lymphatic vessels appears to occur both locally and, in the tumor-draining lymph nodes, it is currently not possible to determine the relative importance of a process occurring at either site for tumor progression. The first sites where tumor-specific immune cells respond to tumor antigens are the draining lymph nodes. This occurs in the early stages of most cancers [24]. Sentinel LNs have attracted more attention because they are the first LNs to drain the tumor bed and are therefore expected to be the first site of tumor metastasis [25]. The presence of tumor cells in lymph nodes is considered evidence of the escape of the tumor from immune surveillance. Lymph nodes are specialized structures for the development of either cell-mediated or humoral immune responses [26].

The existence of tumor-specific effector T cells and immunosuppressive cells has been shown in tumor-draining lymph nodes [27]. The delicate balance between effector and regulatory T cells is suggested to be important in determining the outcome of immune responses to tumors [28]. On the other hand, in mouse models of cancer, B cells accounted for one-third of the lymphocyte population in tumor-draining lymph nodes [29], suggesting that B cells have crucial roles in the formation and/or modulation of anti-tumor immunity.

The three steps of immune surveillance – elimination, equilibrium, and escape – also take place in tumor draining lymph nodes (TDLNs) with LN involvement, which typically represents the escape route for the tumor [30]. Therefore, TDLNs are strategic for both anti-tumor responses and tumor metastasis. An important and still unanswered question is how an arsenal of immune cells can be so overwhelmed by tumor cells that they are eventually completely overpowered by them and become a major pathway for their spread. By analyzing the cellular composition and immune responses at different stages of cancer progression, researchers have sought to answer this question. Indeed, they found that when exposed to tumor products or tumor invasion, TDLNs undergo structural and cellular changes, usually favoring further tumor progression [31–34]. According to this, investigation of the immune status of TDLN may help us understand the mechanism of immunosuppression associated with cancer patients. Several researchers studied the "immunomorphology" of draining lymph nodes of tumors by using hematoxylin and eosin staining and their association with disease parameters and outcomes [35, 36]. For example, head and neck cancer patients who have lymphocytic predominance in the draining lymph nodes have a lower risk of metastatic disease, whereas those who have germinal center predominance have a higher risk of metastatic disease [36, 37].

In addition, flow cytometry data showed a significant decrease in CD8+ T cell frequency in lymph nodes from head and neck cancer patients compared with control LNs [38]. The mechanisms responsible for this change may be complicated; however, the immunosuppressive effects mediated by tumor-associated factors, cytokines, and selective depletion of CD8+ T cells as a result of chronic antigenic stimulation might be important [38, 39]. In fact, suppression of the immune responses within TDLNs is a critical step for nodal invasion [27, 40]. It was shown that immunosuppressive events occur in regional LNs in melanoma even before nodal metastasis [41]. A comparison of the frequency of CD8+ T cells in TDLNs of melanoma patients with dormant and infectious inflamed LNs revealed that CD8+ T cells are decreased in both metastatic and nonmetastatic LNs. Furthermore, no difference was found in the expression of CD4, CD8, CD14, CD40, CD86, CD123, HLA-DR, and IL-10 in metastatic and nonmetastatic LNs of melanoma patients [41]. A similar observation was reported in breast cancer; CD4+ and CD8+ T cells were significantly decreased in sentinel (SLNs) and axillary (ALNs) lymph nodes of breast cancer patients compared to controls. A decrease in CD1a DCs in SLNs was also reported. The interesting finding was that the frequency of CD4+ T cells was reduced even in nonmetastatic ALNs, further supporting the idea that changes in the immune profile of TDLNs are dynamic and may precede nodal involvement. Another finding of the study was the association between the percentage of CD4+

T cells and DCs in ALNs and patients' disease-free survival, independent of nodal involvement [42].

In situations of chronic inflammation, organized lymphoid structures containing DCs, T cells, and B cells are formed in nonlymphoid tissues, called tertiary lymphoid structures (TLS), similar to those seen in secondary lymphoid tissues [43]. The function of these TLS is probably to provide primary and local defense against microbes, and their constituent cells will disappear once the pathogen has been eliminated. TLS is seen in autoimmune and infectious diseases, graft rejection, and cancer at the site of inflammation. However, TLS has been shown to be a feature of autoimmune disease, where the large number of adaptive immune cells in these structures can exacerbate autoimmune disease [44]. On the other hand, TLS may play a role in cancer progression in the context of chronic inflammation, but it has been suggested that adaptive anti-tumor immune responses can be generated in TLS [45, 46]. In nonsmall cell lung cancer (NSCLC), TLSs are found in tumor tissues and are composed of mature DCs and T cells which are located adjacent to follicles containing GC B cells and FDCs. DC density in these TLS was reported to be a positive prognostic indicator in NSCLC patients [47]. TLSs are reported to be present in tumor tissues of breast cancer and the presence of Tfh cells in TLSs has been associated with better disease outcome. However, a recent study showed that breast cancer tissues usually contained TLS, but the presence of TLS was associated with high grade and aggressive form of the tumor [48].

Lymphangiogenesis process could be targeted by monoclonal antibodies (mAbs) including bevacizumab (anti-VEGF antibody), cediranib (anti-VEGFR antibody) alone, or in combination with kinase inhibitors. However, some clinical trials of anti-lymphangiogenic therapy for solid tumors failed to show an advantage in these patients [49]. In addition to lymphangiogenesis inhibitors, other approaches have been proposed in an attempt to target the lymphatic system. Given that TDLNs play a key role in generating an anti-tumor immune response and have also been shown to be important in suppressing tumor immunity, the most effective treatments are those that aim to tip the balance toward more effective T-cell responses. Further in vitro and in vivo studies are needed to determine whether immunotherapeutic approaches such as tumor-specific T-cell activation, elimination of immunosuppressive pathways, or a combination of these approaches have a blocking effect on tumor growth and metastasis.

Some strategies target tumor or tumor-draining LNs to circumvent tumor-induced immune suppression. Immunostimulatory strategies such as targeting CD40, Toll-like receptor (TLR) ligands, CTLA-4, PD-1 or using inflammatory or pro-inflammatory cytokines (IL-12, TNF-α, IFN-α or IL-2) have been used systemically in experimental models and clinical trials [50, 51]. Preclinical studies and clinical trials have shown that most of these therapies can enhance innate and/or adaptive immunity against tumors [52]. However, systemic approaches have several side effects, including severe toxicity due to systemic activation of the immune system [53].

2.2 Autoimmunity

The lymphatic system has not been the subject of much research about autoimmune diseases. The lymphatic system is a network of low-pressure vessels that provide a pathway for intercellular fluid to return to the blood vessel network. There is a network of lymphatic vessels throughout the body. In addition to returning intercellular fluid to the circulatory system, the lymphatic system also performs important immune functions in the body to keep tissues healthy and functioning properly [2]. The lymphatic vasculature is more conducive to immune induction and tolerance than the blood vasculature in peripheral tissues, which draws leukocytes from contaminated sites for effector functions. The lymphatic vessels are considered an important element of the immune system, not only in maintaining tissue fluid homeostasis but also in transporting antigens from the periphery to the lymph nodes, where lymphocytes are activated, expanded and ultimately transported to the site of inflammation or infection. In addition to the transport of lymphoid and immune cells, the lymphoid system is directly involved in the regulation of the immune system and the induction of tolerance to self-antigens [2, 54].

The ability of LECs to influence the activity of immune cells through a variety of mechanisms has been clearly demonstrated in recent studies. LECs secrete a wide range of cytokines to regulate the immune system. LEC-derived TGF-β has been reported to be a suppressor of dendritic cell maturation [55]. Increased IL-7 production by LECs can lead to the expansion of regulatory T cells, enhancing their immunoregulatory function [56]. LECs direct lymphocytes and DCs to infiltrate or exit the lymph nodes, while inflammation increases their ability to recruit cells. Through high levels of PD-L1 expression and their lack of costimulatory molecules, LEC also ensures that CD8 T cells are able to tolerate peripheral tissue antigens. Understanding how T-cell fate is influenced by other inhibitory molecules expressed in LEC will be extremely interesting. In addition, knowing whether LEC is able to induce CD4 T-cell tolerance or serves as a reservoir of peripheral tissue antigen in DC for presentation will provide more certainty about the general immunoregulatory role of LEC [57].

Alterations in the lymphatic system have also been observed in most autoimmune diseases, particularly in terms of lymphatic vessel phenotype and patterns of denovo lymphangiogenesis. For instance, in psoriasis, an autoimmune and inflammatory skin condition, many changes in the peripheral lymphatic system have been shown to play an important role in causing the disease to develop [58]. Interestingly, lymphatic vessels have shown the ability to modulate their expression of key immune mediators in response to most cytokines involved in the pathogenesis of psoriasis, including IL-27 or TNF-α [59, 60].

In the case of rheumatic autoimmune diseases, there is some evidence to suggest that lymphatic dysfunction may be a contributing factor in the development of these diseases [61]. This is due to the role of the lymphatic system in the immune system. One of the most studied autoimmune diseases regarding the role of the lymphatic system is rheumatoid arthritis (RA). RA is a chronic inflammatory disease. It causes pain, swelling, and stiffness (reduced flexibility) in the joints. It is a type of arthritis that occurs when the body's immune system mistakenly attacks our joints, destroying and inflaming them [62]. The local lymphatic system is said to undergo two stages of change in association with the general inflammation of rheumatoid arthritis. In response to early rheumatoid arthritis or synovitis, lymphoid tissues undergo an "enlargement" phase, which increases their capacity to remove cellular debris and inflammatory cells from the site of infection, either through lymphangiogenesis or through increased vascular contraction frequency. During this expansion phase, in addition to changes in the lymphatic vessels, the draining lymph nodes themselves also expand characterized by a high infiltration of IgM+ CD23+ CD21hiCD1dhi B cells [63, 64]. This enlarged lymph node in the popliteal area may be a useful marker for the detection of arthritis in the early stages of disease activity. Based on RA model studies, an acute arthritis flare in the early stages of the disease has been observed, with increased lymphatic drainage from inflamed joints to enlarged draining lymph nodes. After a prolonged period of expansion, a stochastic event leads to the asymmetric collapse of LNs and lymphatics. In the collapsed phase, the local lymphatic system collapses, causing a loss of lymphatic flow and a reduction in lymphatic clearance. Coordinated with the collapse, B cells migrate from the follicle into the sinuses. B cell depletion therapy reduces arthritis flares by eliminating these B cells and improving passive lymphatic drainage from inflamed joints [65, 66].

Lupus is a type of autoimmune disease in which the body's immune system attacks its tissues and organs. These attacks cause inflammation, swelling, and damage to different parts of the body, such as the joints, skin, kidneys, blood, heart, and lungs [67]. It is not well understood how the lymphatic system is related to lupus. However, the lymphatic vessels may expand and contract more easily than normal when an attempt is made to remove inflammatory cells in lupus, just as they do in other autoimmune diseases such as rheumatoid arthritis or scleroderma. The lymph nodes also enlarge due to the accumulation and filtration of lymph, resulting in the formation of inflammatory cells and fluid. This can lead to an increase in joint swelling and pain as the fluid volume builds up [68, 69]. Although the nodes do not always become enlarged, they can become swollen during periods of high disease activity or lupus flares. There have been cases of lymph leakage from the lymphatic system into the abdomen and upper spine in people with lupus. Lymphatic blockage or other abnormalities in the flow of lymph fluid may be the cause [68–70]. More needs to be known about what is going on here. The data show that lymphatic dysfunction is a factor contributing to photosensitivity in murine lupus and can even be alleviated by improving lymphatic flow with manual lymphatic drainage (MLD) [68]. The causes of lymphoedema in murine lupus and the mechanisms for reduced photosensitivity through improved lymphatic drainage will be investigated in future studies. Altering the lymphatic system may be a novel target for new drugs if similar immune circuit defects occur in patients with SLE.

Scleroderma is a chronic autoimmune disease that causes thickening and hardening of the skin, scarring, and damage to internal organs such as the heart and blood vessels, lungs, stomach, and kidneys. Scleroderma is caused by the overproduction and accumulation of collagen in the body's tissues [71]. Collagen is a type of protein fiber that makes up the body's connective tissues, including the skin. It is difficult to understand what causes the disease, but defects in vascular and cellular function are thought to be key driving factors. In the study of vascular dysfunction, blood endothelial cells have been highlighted, whereas lymphatic endothelial cells (LECs) have been much less studied. Skin samples from patients with scleroderma have been shown to be indicative of lymphatic dysfunction, based on studies conducted in 1999 by Leu et al. [72]. There has also been evidence of lymphatic changes in other diseases associated with fibrosis [73]. It also suggests that lymphatic dysfunction might be a therapeutic target for scleroderma.

With a better understanding of how the lymph system is affected, it is important to test approaches that can help treat lymph problems. There are several forms of manual therapy and manual lymphatic drainage (MLD) is one of them. Lymphatic massage is one of the most sensitive and important types of massage and is the drainage of lymph. Lymph drainage improves the function of the body's lymphatic system and increases its efficiency. For example, people with scleroderma have edema, which is caused by fluid accumulation, and MLD has been shown to reduce swelling and have an effect on hand function [74]. Likewise, lymphatic dry brushing is recommended to reduce lymphatic congestion by reducing fluid accumulation and inflammation. This 5000-year-old technique is based on ancient Ayurvedic medicine in India [75].

2.3 Inflammation and infection

The lymphatic system removes infections and maintains a balance of fluids in the body. Abnormalities in lymphatic function are the cause of the disease, which is often characterized by reduced lymph flow and swelling in the affected limbs such as lymphedema, which can have immune-deficiency consequences. Infectious diseases are the main cause of acquired lymphoedema [76]. Despite the strong association between infection and lymphoedema, it is not well understood whether the lymphatic system is involved in the pathogenesis of bacterial infections or how it responds to microbial and viral infections.

Infections can also cause other problems with the lymphatic system, which are divided into two groups according to the site of damage: lymphadenitis or lymph-adenopathy, which affects the lymph nodes, and lymphangitis, which affects the lymphatic vessels [77].

The lymph nodes play an important role in the immune system's response to infection. Swollen lymph nodes are usually the result of a bacterial or viral infection. Several factors can cause lymph nodes to swell, including colds, flu, ear infections, tuberculosis, and strep throat [78]. In rare cases, this swelling can be caused by more serious conditions such as lymphoma. The amount of swelling depends on the severity of the infection or inflammation. The more serious the condition, the more swollen and painful the gland will be. Swollen lymph nodes can also be a sign of an autoimmune disease, such as rheumatoid arthritis and lupus, or an unusual infection, such as mononucleosis and AIDS. These diseases involve the immune system in all parts of the body and cause an accumulation of lymphocytes in the lymph nodes [54, 77].

Inflammation or infection has been associated with an increase in the number of dendritic cells that leave the site of disease and enter the afferent lymphatic vessels through the induction of chemokine receptors and adhesion molecules. The lymphatic system uses chemokines and counter receptors to regulate the circulation of immune cells within the lymph node [79]. In the inflamed lymph node, there is an increase in the mobilization of immune cells into the gland and a temporary decrease in lymphocytes, leaving the draining lymph nodes. These inflammatory changes, together with the highly specialized architecture of the lymph nodes, increase the likelihood of antigen presentation to the cognate lymphocyte [80]. During inflammation, the immune cells of the draining lymph node undergo changes in both phenotype and function [81]. In response to TNF-α stimuli, dendritic cells have been shown to upregulate MHC class II and the costimulatory molecules and downregulate the chemokine receptors, which contribute to the recruitment of leukocytes to the tissue [82]. In order to selectively increase the lymphatic trafficking of specific immune cells, the transcriptional profile of the lymphatic endothelial cells may also be altered [83]. In a study of human dermal LECs, LECs were shown to release exosomes that cluster around lymphatic vessels under inflammatory conditions and promote dendritic cell migration to lymph nodes [84]. The interaction of LECs with DCs has been studied, and the results show that cooperation between TNF-stimulated LECs and DCs through αMβ2 integrin: ICAM-1 interaction causes downregulation of the costimulatory molecule CD86 on DCs, which can lead to impairment of dendritic cells in activating T cells (**Figure 2**) [23].

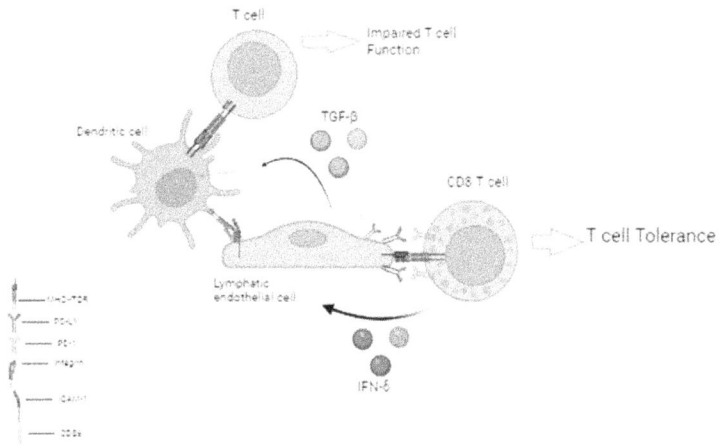

Figure 2.
Schematic showing how lymphatic endothelial cells (LECs) interact with dendritic cells (DCs) and T cells in chronic inflammation (e.g., cancer/infection/autoimmunity). LECs acquire/exchange antigens. Transcription factor gene variations affect lymphatic endothelium phenotype. IFN-γ induces PD-L1 expression in LECs from antigen-specific CD8+ T cells. LECs induce reduced CD86 expression on DCs via TGF-β, impairing T-cell function and promoting immune tolerance.

Changes in lymphatic endothelial cell junction are also induced by inflammation. In endothelial cells of collecting lymphatic vessels, replacement of zipper junctions by button junctions was observed sometime after inflammation-induced mycoplasma infection and treatment with corticosteroids (anti-inflammatory drugs) could reverse these junctional changes [85]. These changes in the inflamed lymphatic vessels can affect the balance between fluid entry and drainage, and the removal of inflammation. However, the flexibility of these junctions makes them a suitable target for reducing chronic inflammation. In a mouse model of mycoplasma pulmonis, both blood and lymphatic vessels have been observed to undergo capillary to venular dilation and lymphangiogenesis. TNF-α has been shown to have a direct effect on both processes, but it appears that the effect on lymphangiogenesis requires further inflammatory mediators from leukocytes [86].

It is clear that lymphangiogenesis and lymphatic remodeling, which are associated with inflammation, are interesting therapeutic targets. An expanded lymphatic network would allow better lymphatic transport. This would presumably improve the removal of excess extracellular fluid caused by inflammation-induced vascular permeability. However, immune surveillance in the lymph nodes could be overwhelmed by trafficking and such proangiogenic signaling pathways need to be carefully controlled.

3. Conclusion

The critical role of the lymphatic system in maintaining tissue homeostasis as well as in disease has been well established over the last decades. Apart from their contribution to immune and inflammatory diseases, a large body of scientific literature documents the importance of lymph vessels for cancer biology and the prognostic value of cancer patients.

From a clinical point of view, targeting the lymphatic system has the potential to improve vaccine delivery and tolerance induction and increase the utility of therapies for a wide range of conditions, from infections like HIV to cancer, inflammation, and metabolism. Much progress could be made with tools that target lymphatic vessels in a specific tissue or lymph nodes, for example, using specific antibodies or mouse models that allow gene deletion at both sites. New insights into organ-specific differences in the lymphatic vasculature, such as those provided by newly developed transcriptomic analyses, may provide new opportunities in the future. More work is needed to better understand the complex interplay of LECs and the immune system, and ultimately, to translate this knowledge into therapeutic applications.

Author details

Marzieh Norouzian[1]* and Soghra Abdi[2]

1 Department of Laboratory Sciences, School of Allied Medical Sciences, Hormozgan University of Medical Sciences, Bandar Abbas, Iran

2 Department of Immunology, School of Medicine, Shiraz University of Medical Sciences, Shiraz, Iran

*Address all correspondence to: marzieh.norouzi@gmail.com

IntechOpen

References

[1] Drayton DL et al. Lymphoid organ development: From ontogeny to neogenesis. Nature Immunology. 2006;**7**(4):344-353

[2] Breslin JW et al. Lymphatic vessel network structure and physiology. Comprehensive Physiology. 2018;**9**(1):207-299

[3] Jalkanen S, Salmi M. Lymphatic endothelial cells of the lymph node. Nature Reviews Immunology. 2020;**20**(9):566-578

[4] Zoltzer H. Initial lymphatics-morphology and function of the endothelial cells. Lymphology. 2003;**36**(1):7-25

[5] Tamburini BAJ et al. Chronic liver disease in humans causes expansion and differentiation of liver lymphatic endothelial cells. Frontiers in Immunology. 2019;**10**:1036

[6] Kataru RP et al. Regulation of immune function by the lymphatic system in lymphedema. Frontiers in Immunology. 2019;**10**:470

[7] Yuan Y et al. Modulation of immunity by lymphatic dysfunction in lymphedema. Frontiers in Immunology. 2019;**10**:76

[8] Lucas ED, Tamburini BAJ. Lymph node lymphatic endothelial cell expansion and contraction and the programming of the immune response. Frontiers in Immunology. 2019;**10**:36

[9] Padera TP, Meijer EF, Munn LL. The lymphatic system in disease processes and cancer progression. Annual Review of Biomedical Engineering. 2016;**18**:125-158

[10] Kim H, Kataru RP, Koh GY. Inflammation-associated lymphangiogenesis: A double-edged sword? The Journal of Clinical Investigation. 2014;**124**(3):936-942

[11] Christiansen A, Detmar M. Lymphangiogenesis and cancer. Genes & Cancer. 2011;**2**(12):1146-1158

[12] Radhakrishnan K, Rockson SG. The clinical spectrum of lymphatic disease. Annals of the New York Academy of Sciences. 2008;**1131**(1):155-184

[13] Margaris K, Black RA. Modelling the lymphatic system: Challenges and opportunities. Journal of the Royal Society Interface. 2012;**9**(69):601-612

[14] Anderson NM, Simon MC. The tumor microenvironment. Current Biology. 2020;**30**(16):R921-r925

[15] Mumprecht V et al. In vivo imaging of inflammation-and tumor-induced lymph node lymphangiogenesis by immuno–positron emission tomography. Cancer Research. 2010;**70**(21):8842-8851

[16] Qian C-N et al. Preparing the "soil": The primary tumor induces vasculature reorganization in the sentinel lymph node before the arrival of metastatic cancer cells. Cancer Research. 2006;**66**(21):10365-10376

[17] Murdoch C et al. The role of myeloid cells in the promotion of tumour angiogenesis. Nature Reviews Cancer. 2008;**8**(8):618-631

[18] Clasper S et al. A novel gene expression profile in lymphatics associated with tumor growth and nodal metastasis. Cancer Research. 2008;**68**(18):7293-7303

[19] Lane RS et al. IFNγ-activated dermal lymphatic vessels inhibit cytotoxic T cells in melanoma and inflamed skin. Journal of Experimental Medicine. 2018;**215**(12):3057-3074

[20] Gkountidi AO et al. MHC class II antigen presentation by lymphatic endothelial cells in tumors promotes intratumoral regulatory T cell–suppressive functions. Cancer Immunology Research. 2021;**9**(7):748-764

[21] Lund AW et al. VEGF-C promotes immune tolerance in B16 melanomas and cross-presentation of tumor antigen by lymph node lymphatics. Cell Reports. 2012;**1**(3):191-199

[22] Lee E, Pandey NB, Popel AS. Lymphatic endothelial cells support tumor growth in breast cancer. Scientific Reports. 2014;**4**(1):5853

[23] Podgrabinska S et al. Inflamed lymphatic endothelium suppresses dendritic cell maturation and function via mac-1/ICAM-1-dependent mechanism. The Journal of Immunology. 2009;**183**(3):1767-1779

[24] Evans EM et al. Infiltration of cervical cancer tissue with human papillomavirus-specific cytotoxic T-lymphocytes. Cancer Research. 1997;**57**(14):2943-2950

[25] Mabry H, Giuliano AE. Sentinel node mapping for breast cancer: Progress to date and prospects for the future. Surgical Oncology Clinics of North America. 2007;**16**(1):55-70

[26] Munn DH, Mellor AL. The tumor-draining lymph node as an immune-privileged site. Immunological Reviews. 2006;**213**(1):146-158

[27] Fransen MF, Arens R, Melief CJ. Local targets for immune therapy to

cancer: Tumor draining lymph nodes and tumor microenvironment. International Journal of Cancer. 2013;**132**(9):1971-1976

[28] Mougiakakos D. Regulatory T cells in colorectal cancer: From biology to prognostic relevance. Cancers. 2011;**3**(2):1708-1731

[29] Li Q et al. Simultaneous targeting of CD3 on T cells and CD40 on B or dendritic cells augments the antitumor reactivity of tumor-primed lymph node cells. The Journal of Immunology. 2005;**175**(3):1424-1432

[30] Swann JB, Smyth MJ. Immune surveillance of tumors. The Journal of Clinical Investigation. 2007;**117**(5):1137-1146

[31] Gai XD et al. Potential role of plasmacytoid dendritic cells for FOXP3+ regulatory T cell development in human colorectal cancer and tumor draining lymph node. Pathology-Research and Practice. 2013;**209**(12):774-778

[32] Shuang Z-Y et al. The tumor-draining lymph nodes are immunosuppressed in patients with hepatocellular carcinoma. Translational Cancer Research. 2017;**6**(6):1188-1196

[33] Munn DH et al. Expression of indoleamine 2, 3-dioxygenase by plasmacytoid dendritic cells in tumor-draining lymph nodes. The Journal of Clinical Investigation. 2004;**114**(2):280-290

[34] Norouzian M et al. Regulatory and effector T cell subsets in tumor-draining lymph nodes of patients with squamous cell carcinoma of head and neck. BMC Immunology. 2022;**23**(1):56

[35] Cernea C et al. Prognostic significance of lymph node reactivity in the control of pathologic negative node

squamous cell carcinomas of the oral cavity. The American Journal of Surgery. 1997;**174**(5):548-551

[36] Chandavarkar V et al. Immunomorphological patterns of cervical lymph nodes in oral squamous cell carcinoma. Journal of Oral and Maxillofacial Pathology: JOMFP. 2014;**18**(3):349

[37] Yadav ST et al. Immunomor phological assessment of regional lymph nodes for predicting metastases in oral squamous cell carcinoma. Indian Journal of Dental Research. 2012;**23**(1):121

[38] Lapointe H, Lampe H, Banerjee D. Head and neck squamous cell carcinoma cell line-induced suppression of in vitro lymphocyte proliferative responses. Otolaryngology–Head and Neck Surgery. 1992;**106**(2):149-158

[39] Akbar AN et al. The significance of low bcl-2 expression by CD45RO T cells in normal individuals and patients with acute viral infections. The role of apoptosis in T cell memory. The Journal of Experimental Medicine. 1993;**178**(2):427-438

[40] Hood JL, San RS, Wickline SA. Exosomes released by melanoma cells prepare sentinel lymph nodes for tumor metastasis. Cancer Research. 2011;**71**(11):3792-3801

[41] Mansfield AS et al. Regional immunity in melanoma: immunosuppressive changes precede nodal metastasis. Modern Pathology. 2011;**24**(4):487-494

[42] Kohrt HE et al. Profile of immune cells in axillary lymph nodes predicts disease-free survival in breast cancer. PLoS Medicine. 2005;**2**(9):e284

[43] Chen J, Chen J, Wang L. Tertiary lymphoid structures as unique constructions associated with the organization, education, and function of tumor-infiltrating immunocytes. Journal of Zhejiang University. Science. B. 2022;**23**(10):812-822

[44] Shipman WD, Dasoveanu DC, Lu TT. Tertiary lymphoid organs in systemic autoimmune diseases: Pathogenic or protective? F1000Res. 2017;**6**:196

[45] Germain C, Gnjatic S, Dieu-Nosjean M-C. Tertiary lymphoid structure-associated B cells are key players in anti-tumor immunity. Frontiers in Immunology. 2015;**6**:67

[46] Dieu-Nosjean M-C et al. Tertiary lymphoid structures in cancer and beyond. Trends in Immunology. 2014;**35**(11):571-580

[47] Dieu-Nosjean M-C et al. Long-term survival for patients with non–small-cell lung cancer with intratumoral lymphoid structures. Journal of Clinical Oncology. 2008;**26**(27):4410-4417

[48] Figenschau SL et al. Tertiary lymphoid structures are associated with higher tumor grade in primary operable breast cancer patients. BMC Cancer. 2015;**15**:1-11

[49] Padera TP et al. Differential response of primary tumor versus lymphatic metastasis to VEGFR-2 and VEGFR-3 kinase inhibitors cediranib and vandetanib. Molecular Cancer Therapeutics. 2008;**7**(8):2272-2279

[50] Melero I et al. Immunostimulatory monoclonal antibodies for cancer therapy. Nature Reviews Cancer. 2007;**7**(2):95-106

[51] Vacchelli E et al. Trial watch: Immunostimulatory cytokines. Oncoimmunology. 2012;**1**(4):493-506

[52] Aranda F et al. Trial watch: Immunostimulatory monoclonal antibodies in cancer therapy. Oncoimmunology. 2014;**3**(2):e27297

[53] Stucci S et al. Immune-related adverse events during anticancer immunotherapy: Pathogenesis and management. Oncology Letters. 2017;**14**(5):5671-5680

[54] Grant SM et al. The lymph node at a glance - how spatial organization optimizes the immune response. Journal of Cell Science. 2020;**133**(5):jcs241828

[55] Christiansen AJ et al. Lymphatic endothelial cells attenuate inflammation via suppression of dendritic cell maturation. Oncotarget. 2016;**7**(26):39421

[56] Schmaler M et al. IL-7R signaling in regulatory T cells maintains peripheral and allograft tolerance in mice. Proceedings of the National Academy of Sciences. 2015;**112**(43):13330-13335

[57] Dieterich LC et al. Tumor-associated lymphatic vessels upregulate PDL1 to inhibit T-cell activation. Frontiers in Immunology. 2017;**8**:66

[58] Henno A et al. Altered expression of angiogenesis and lymphangiogenesis markers in the uninvolved skin of plaque-type psoriasis. British Journal of Dermatology. 2009;**160**(3):581-590

[59] Shibata S et al. Possible roles of IL-27 in the pathogenesis of psoriasis. Journal of Investigative Dermatology. 2010;**130**(4):1034-1039

[60] Coimbra S et al. The roles of cells and cytokines in the pathogenesis of psoriasis. International Journal of Dermatology. 2012;**51**(4):389-398

[61] Schwartz N et al. Lymphatic function in autoimmune diseases. Frontiers in Immunology. 2019;**10**:519

[62] Radu A-F, Bungau SG. Management of rheumatoid arthritis: An overview. Cell. 2021;**10**(11):2857

[63] Bouta EM et al. The role of the lymphatic system in inflammatory-erosive arthritis. Seminars in Cell & Developmental Biology. 2015;**38**:90-97

[64] Rahimi H et al. Lymphatic imaging to assess rheumatoid flare: Mechanistic insights and biomarker potential. Arthritis Research & Therapy. 2016;**18**(1):194

[65] Li J et al. Expanded CD23+/CD21hi B cells in inflamed lymph nodes are associated with the onset of inflammatory-erosive arthritis in TNF-transgenic mice and are targets of anti-CD20 therapy. The Journal of Immunology. 2010;**184**(11):6142-6150

[66] Li J et al. Efficacy of B cell depletion therapy for murine joint arthritis flare is associated with increased lymphatic flow. Arthritis and Rheumatism. 2013;**65**(1):130-138

[67] Mok C, Lau C. Pathogenesis of systemic lupus erythematosus. Journal of Clinical Pathology. 2003;**56**(7):481-490

[68] Ambler WG et al. 205 lymphatic dysfunction in lupus photosensitivity. Archives of Disease in Childhood. 2021;**8**(Suppl 2):A1-A75

[69] Rajasekhar L et al. Lymphatic obstruction as a cause of extremity edema in systemic lupus erythematosus. Clinical Rheumatology. 2013;**32**:11-13

[70] Daniel A et al. Chylous ascites in a patient with an overlap syndrome: A surprising response to rituximab. Case Reports. 2017;**2017**:bcr-2017-222339

[71] Gabrielli A, Avvedimento EV, Krieg T. Scleroderma. New

England Journal of Medicine. 2009;**360**(19):1989-2003

[72] Leu A et al. Lymphatic microangiopathy of the skin in systemic sclerosis. Rheumatology (Oxford, England). 1999;**38**(3):221-227

[73] Garber S et al. Enlarged mediastinal lymph nodes in the fibrosing alveolitis of systemic sclerosis. The British Journal of Radiology. 1992;**65**(779):983-986

[74] Bongi SM et al. Manual lymph drainage improving upper extremity edema and hand function in patients with systemic sclerosis in edematous phase. Arthritis Care & Research. 2011;**63**(8):1134-1141

[75] Narahari S et al. Integrated management of filarial lymphedema for rural communities. Lymphology. 2007;**40**(1):3-13

[76] Sleigh BC, Manna B. Lymphedema. Treasure Island (FL): StatPearls Publishing; 2019

[77] Zeppa P, Cozzolino I. Lymphadenitis and lymphadenopathy. Lymph Node FNC. 2018;**23**:19-33

[78] Sahai S. Lymphadenopathy. Pediatrics in Review. 2013;**34**(5):216-227

[79] Neeland MR, Meeusen ENT, de Veer MJ. Afferent lymphatic cannulation as a model system to study innate immune responses to infection and vaccination. Veterinary Immunology and Immunopathology. 2014;**158**(1):86-97

[80] Tan KW et al. Expansion of cortical and medullary sinuses restrains lymph node hypertrophy during prolonged inflammation. The Journal of Immunology. 2012;**188**(8):4065-4080

[81] Johnson LA et al. An inflammation-induced mechanism for leukocyte transmigration across lymphatic vessel endothelium. The Journal of Experimental Medicine. 2006;**203**(12):2763-2777

[82] Sallusto F et al. Rapid and coordinated switch in chemokine receptor expression during dendritic cell maturation. European Journal of Immunology. 1998;**28**(9):2760-2769

[83] Vigl B et al. Tissue inflammation modulates gene expression of lymphatic endothelial cells and dendritic cell migration in a stimulus-dependent manner. Blood, The Journal of the American Society of Hematology. 2011;**118**(1):205-215

[84] Brown M et al. Lymphatic exosomes promote dendritic cell migration along guidance cues. The Journal of Cell Biology. 2018;**217**(6):2205-2221

[85] Baluk P et al. Functionally specialized junctions between endothelial cells of lymphatic vessels. The Journal of Experimental Medicine. 2007;**204**(10):2349-2362

[86] Baluk P et al. TNF-α drives remodeling of blood vessels and lymphatics in sustained airway inflammation in mice. The Journal of Clinical Investigation. 2009;**119**(10):2954-2964

www.ingramcontent.com/pod-product-compliance
Lightning Source LLC
Chambersburg PA
CBHW081237190326
41458CB00016B/5817